THE CONSTRUCTION LAW LIBRARY FROM WILEY LAW PUBLICATIONS

Alternative Clauses to Standard Construction Contracts
James E. Stephenson, Editor

Alternative Dispute Resolution in the Construction Industry
Robert F. Cushman, G. Christian Hedemann, and Avram S. Tucker, Editors

Arbitration of Construction Disputes
Michael T. Callahan, Barry B. Bramble, and Paul M. Lurie

Architect and Engineer Liability: Claims against Design Professionals
Robert F. Cushman and Thomas G. Bottum, Editors

Calculating Construction Damages
William Schwartzkopf, John J. McNamara, and Julian F. Hoffar

California Construction Defect Litigation: Residential and Commercial (second edition)
Thomas E. Miller

California Construction Law (fifteenth edition)
Kenneth C. Gibbs and Gordon Hunt

Construction and Environmental Insurance Case Digests
Wiley Law Publications, Editors

Construction Bidding Law
Robert F. Cushman and William J. Doyle, Editors

Construction Claims and Liability
Michael S. Simon

Construction Contracting Alert Newsletter
Smith, Currie & Hancock

Construction Contractor's Handbook of Business and Law
Robert F. Cushman, G. Christian Hedemann, and Peter J. King, Editors

Construction Defaults: Rights, Duties, and Liabilities
Robert F. Cushman and Charles A. Meeker, Editors

Construction Delay Claims (second edition)
Barry B. Bramble and Michael T. Callahan

Construction Documentation (second edition)
John A. Ricchini and James J. O'Brien

Construction Engineering Evidence
Loren W. Peters

Construction Industry Contracts: Legal Citator and Case Digest
Wiley Law Publications Editorial Staff

Construction Industry Forms (two volumes)
Robert F. Cushman and George L. Blick, Editors

Construction Industry Insurance Handbook
Deutsch, Kerrigan & Stiles

Construction Litigation Formbook
David M. Buoncristiani, John D. Carter, and Robert F. Cushman

Construction Litigation: Representing the Contractor (second edition)
John D. Carter, Robert F. Cushman, and William J. Palmer, Editors

Construction Litigation: Representing the Owner (second edition)
Robert F. Cushman, Kenneth M. Cushman, and Stephen B. Cook, Editors

Construction Litigation: Strategies and Techniques
Barry B. Bramble and Albert E. Phillips, Editors

Construction Renovation Formbook
Robert F. Cushman and John W. DiNicola, Editors

Construction Scheduling: Preparation, Liability, and Claims
Jon M. Wickwire, Thomas J. Driscoll, and Stephen B. Hurlbut

Construction Subcontracting: A Legal Guide for Industry Professionals
Overton A. Currie, Neal J. Sweeney, and Randall F. Hafer, Editors

Construction Surety and Bonding Handbook
Gorden Hunt

Design-Build Contracting Handbook
Robert F. Cushman and Kathy Sperling Taub, Editors

Design Professional's Handbook of Business and Law
Robert F. Cushman and James C. Dobbs, Editors

Differing Site Condition Claims
Robert F. Cushman and David Tortorello, Editors

Federal Construction Contracting
James F. Nagle

Fifty State Construction Lien and Bond Law (two volumes)
Robert F. Cushman, Editor

Hazardous Waste Disposal and Underground Construction Law
Robert F. Cushman and Bruce W. Ficken, Editors

Legal Guide to AIA Documents (third edition)
Werner Sabo

1992 Directory of Construction Industry Consultants
Wiley Law Publications, Editors

1992 Wiley Construction Law Update
Overton A. Currie and Neal J. Sweeney, Editors

Proving and Pricing Construction Claims
Robert F. Cushman and David A. Carpenter, Editors

Sweet on Construction Industry Contracts (second edition)
Justin Sweet

Troubled Construction Loans: Law and Practice
Stanley P. Sklar, Editor

DIFFERING SITE CONDITION CLAIMS

SUBSCRIPTION NOTICE

This Wiley product is updated on a periodic basis with supplements to reflect important changes in the subject matter. If you purchased this product directly from John Wiley & Sons, Inc., we have already recorded your subscription for this update service.

If, however, you purchased this product from a bookstore and wish to receive (1) the current update at no additional charge, and (2) future updates and revised or related volumes billed separately with a 30-day examination review, please send your name, company name (if applicable), address, and the title of the product to:

Supplement Department
John Wiley & Sons, Inc.
One Wiley Drive
Somerset, NJ 08875
1-800-225-5945

DIFFERING SITE CONDITION CLAIMS

Robert F. Cushman

David R. Tortorello

Editors

Wiley Law Publications

JOHN WILEY & SONS, INC.

New York • Chichester • Brisbane • Toronto • Singapore

In recognition of the importance of preserving what has been written, it is a policy of John Wiley & Sons, Inc., to have books of enduring value published in the United States printed on acid-free paper, and we exert our best efforts to that end.

This publication is designed to provide accurate and authoritative information in regard to the subject matter covered. It is sold with the understanding that the publisher is not engaged in rendering legal, accounting, or other professional services. If legal advice or other expert assistance is required, the services of a competent professional person should be sought. *From a Declaration of Principles jointly adopted by a Committee of the American Bar Association and a Committee of Publishers.*

Library of Congress Cataloging-in-Publication Data

Differing site condition claims / Robert F. Cushman, David R. Tortorello, editors.
p. cm. — (Construction law library)
ISBN 0-471-55203-8 (cloth)
1. Construction contracts—United States. I. Cushman, Robert Frank, 1931– . II. Tortorello, David R. III. Series.
KF902.D53 1992
343.73'07869—dc20
[347.3037869] 92-21038
CIP

Printed in the United States of America

10 9 8 7 6 5 4 3 2 1

PREFACE

A *differing site condition* (also called a *changed condition*) is a physical condition other than the weather, climate, or another act of God discovered on or affecting a construction site that differs in some material respect from what reasonably was anticipated. For example, encountering rock instead of sand, or substantially more rock than indicated, or pipes wrapped with asbestos, or drums of a hazardous substance are conditions in the field that differ from those expected.

The difference often results from inaccuracies in the owner's plans and specifications. For instance, soil conditions may have been inaccurately portrayed or buried utility lines may not have been properly located on the drawings. Sometimes the difference results from the severe or unusual nature of the condition itself. If the physical condition could not have been foreseen by an experienced, prudent contractor, it also can be considered a differing site condition.

Most construction contracts, both public and private, contain a differing site condition clause authorizing additional compensation to cover the contractor's cost of coping with differing conditions. The rationale is that if the owner agrees to share this risk, contractors will not be forced to carry large contingencies in their bids and prices will be more competitive.

Given the physical uncertainties inherent in any construction project and the widespread use of differing site condition clauses, it is not surprising that a large number of these claims occur. The successful prosecution of such claims requires knowledge of the law in this area as well as skilled contract administration. Unfortunately, many contractors and their lawyers unnecessarily bungle legitimate claims.

This is a practical handbook for lawyers, contractors, consultants, and others who prepare, present, and defend differing site condition claims as well as related claims against owners, contractors, and design professionals. It covers every type and aspect of differing site condition claims and clarifies issues of entitlement as well as factual and legal causation.

This book is intended to serve as a reference work for those who must identify claims and preserve their right to pursue them. It is a pathway to those who must determine the remedy-granting clauses in their public or private contracts or are obliged to develop (or defend against) theories of recovery outside of the contracts. It is a road map to the methods of both calculating and proving the amount of damages sustained with reasonable certainty. Finally, it is a practical development of what evidence should be presented in the litigation forum and how best to do so.

The material in this book is presented in a manner that is both instructive and readable to those whose knowledge ranges from slight to extensive. It is a strategic guide to the practicing lawyer who sues or defends as well as a legal treatise that includes recent developments in case law.

The coauthors have been carefully selected from among a group of construction litigators and claims specialists with years of experience in prosecuting and defending differing site condition claims. The breadth and depth of their vast experience becomes most apparent as they develop the key issues involved in proving and pricing these claims. They concisely, yet thoroughly, identify and explain solutions to commonly faced problems.

We believe that through the efforts of these coauthors we have a handbook that will be invaluable to the construction and legal communities. We are indebted to the coauthors for their thorough work, and extend our thanks for their willingness to share their stores of knowledge and experience.

We are confident that this book is a most useful and definitive work for which there has long been a great need.

August 1992

ROBERT F. CUSHMAN
Philadelphia, Pennsylvania

DAVID R. TORTORELLO
New York, New York

ABOUT THE EDITORS

Robert F. Cushman is a partner in the international law firm of Pepper, Hamilton & Scheetz in Philadelphia, Pennsylvania, and is a recognized specialist and lecturer on all phases of construction and real estate law. He serves as legal counsel to numerous trade associations and construction, development, and bonding companies. Mr. Cushman is the editor and coauthor of several books about construction, including the following John Wiley & Sons publications: *Construction Litigation: Representing the Owner (2d ed.); Construction Defaults; Rights, Duties, and Liabilities; Construction Failures; Proving and Pricing Construction Claims;* and *Alternative Dispute Resolution in the Construction Industry.* A member of the Pennsylvania bar, he is admitted to practice before the United States Supreme Court and the Court of Appeals for the Federal Circuit. Mr. Cushman has served as executive vice president and general counsel to the Construction Industry Foundation, as counsel to the American Construction Owners Association, and as regional chairman of the Public Contract Law Section of the American Bar Association. He is permanent chairman of the Andrews Conference Group's Construction Litigation and Hazardous Waste superconferences, and is a charter member of the American College of Construction Lawyers.

David R. Tortorello is the vice president of Peterson Consulting Limited Partnership, responsible for the eastern region construction practice. He provides consulting services for construction projects of all sizes and has substantial experience in construction cost and schedule claim analysis and lost profit analysis. He received his B.S. degree in mechanical engineering from Villanova University in Pennsylvania and his M.B.A. in finance from Pace University, New York, New York. Mr. Tortorello is a professional engineer in New York and is a member of the Project Management Institute and the Construction Financial Management Association. He has presented papers on claims prevention, monitoring, and resolution to various industry groups, and he authored a chapter in *Alternate Dispute Resolution in the Construction Industry* (John Wiley & Sons 1991).

SUMMARY CONTENTS

DETAILED CONTENTS

CHAPTER 1

AVOIDING AND MANAGING RISK OF DIFFERING SITE CONDITIONS

Howard W. Ashcraft, Jr.

Howard W. Ashcraft, Jr. is a partner in Severson & Werson's Construction and Professional Liability Group specializing in litigation and professional practice matters. He has written numerous articles on construction, has lectured nationally on construction law topics and is a former Chairman of the Contract Documents Division of the American Bar Association's Forum on the Construction Industry.

§ 1.1 Introduction

This chapter analyzes site selection and contracting practices as techniques for managing the risk of differing site conditions claims. Site selection and investigation reduce the probability of the risk occurring. Contracting practices transfer the risk to the party best suited to avoid, manage, or bear the risk and reduce the potential impact of the differing site conditions claims.

§ 1.2 Site Selection

Before commencing a project, an owner should be assured that the site is the best available, that it has been adequately investigated, and that the risk of differing site conditions has been reduced to an acceptable level. Compared to the losses that can be sustained in a differing site conditions claim, the cost of adequate investigation is relatively low.

Contractual shifting of risk should be viewed as a secondary or adjunct method of managing the differing site conditions risk. As a practical matter, few contractors will willingly absorb substantial losses due to differing site conditions regardless of the contract terms. If the cost of completion is excessive, litigation may be the contractor's only viable alternative. Even if the owner prevails, the owner will have expended significant sums in litigation costs and may well have lost any opportunity to complete the project in a timely and efficient manner. Avoidance through adequate site selection is always preferable to coping with a differing site conditions claim.

Designers and contractors view site conditions from different perspectives. To designers, site conditions are an engineering parameter for the finished design. In contrast, contractors are interested in the process of construction. Their concern is how the site conditions affect their operations. For example, the depth of groundwater may be irrelevant to the designer if it will not affect foundations but may be very relevant to the contractor if it affects trenching or grading operations.

In most instances, the designers dictate the type of site investigation that will be performed, limiting it to developing information necessary for the final design. Designers scrupulously avoid providing information concerning constructability, and many owners follow suit. Although this may place the designer and the owner in an advantageous position to defend against differing site conditions claims, it probably increases the number of claims asserted because contractors are often misled by reports prepared solely for design.

Consideration of constructability during the site investigation is a bold approach to managing differing site conditions. It increases the potential for the contractor to claim reliance on owner-provided data, but it decreases the probability of a claim. It also permits the owner to effectively resolve problems caused by the differing site condition. Because the constructability problem is discovered before the construction contract is executed, different alternatives can be analyzed or a decision can be made to abandon the project before incurring additional costs. A line on paper is easier to move than a pipeline in the ground.

Site investigation is particularly important when hazardous materials may be encountered. Many construction agreements require the cessation of work should hazardous materials be encountered,[1] and rarely will the construction contractor be equipped to handle the materials found. Because remediation of

---

[1] *See, e.g.,* Engineers Joint Contract Documents Committee 1910-8 (1990), ¶¶ 4.5.1–4.5.3; American Institute of Architects AIA A201-1987, ¶¶ 10.1.2–10.1.3 (asbestos and polychlorinated biphenyls (PCBs)).

hazardous materials requires lengthy investigation, site characterization, and negotiation with governmental entities, the discovery of unforeseen hazardous materials can devastate a construction project.

During the planning stages of a project, the owner should evaluate the necessity of an environmental site assessment. An investigation should be made into the prior uses of the property to assess the probability of encountering underground tanks, pipelines, or hazardous materials that have been improperly stored, discharged, or disposed of. If any of these factors exist, a thorough environmental assessment must be made before proceeding.

Similar problems exist in renovation projects. Often, previously acceptable building materials, such as asbestos, will be encountered during renovation of existing structures. If these are not identified before bid, a differing site condition claim will surely be made.

Early discovery of the differing site condition will reduce the cost of construction. If the problem is disclosed before the bid, the owner obtains a realistic, competitive price for performing the work. If the problem is first discovered during construction, however, the contractor will have little incentive to bargain. Faced with mounting delay and impact costs and the impracticability of changing contractors, the owner may be forced to accept an inflated price for performing the work.

The prudent owner should insist that site investigations consider conditions that could pose difficulties during construction as a means of avoiding differing site conditions claims. At the very least, owners should assure that bidding contractors have sufficient time and adequate opportunity to perform additional investigations.

In some instances, the owner cannot identify and investigate all of the possible sources of differing site conditions. Consideration must then be given to allocating these unknown risks appropriately. Contract provisions are the primary means of accomplishing this result.

§ 1.3 Contract Provisions

A properly drafted construction agreement manages differing site conditions in three distinct ways. First, liability for differing site conditions is dealt with directly. Differing site conditions, site inspection, and reliance clauses are examples of direct clauses. Second, the consequences of a differing site condition can be dealt with indirectly. No-damages-for-delay, per diem delay, unit price, and termination-for-convenience clauses are examples of indirect clauses that limit the owner's liability for cost overruns and delay damages that accompany a differing site conditions claim. Finally, procedural clauses seek to reduce the cost of resolving the differing site conditions claim. Mediation, arbitration, and dispute review board clauses are examples of procedural clauses. Although such clauses are not explicitly concerned with differing site conditions, they may substantially reduce the impact of the differing site condition by reducing the delay and expense involved in resolving the claim.

§ 1.4 —Direct Contract Provisions

Differing Site Conditions Clauses

Differing site conditions clauses seek to equitably allocate an unknown risk between owner and contractor. In theory, this equitable apportionment should minimize costs to the owner because it allows the contractor to remove this contingency from the bid. The owner avoids overpayment on the majority of projects and is required to pay for differing site conditions only when they occur. Utility theory predicts that costs rise when risk is inappropriately allocated,[2] and preliminary studies by the United States Corps of Engineers indicate, in the aggregate, that differing site conditions clauses do reduce the cost of construction.

Nonetheless, there are valid reasons for transferring the risk of unknown site conditions to the contractor. First, the cost efficiencies of a differing site conditions clause are only seen in the aggregate. On any particular project, the efficiency of the clause will depend upon the conditions actually encountered. An owner that rarely engages in construction may be more concerned with the potential for a catastrophic cost overrun than the incrementally higher construction costs that the differing site conditions clause may cause. Second, some owners, particularly public owners, have limited funds for the construction of a project. Substantially increasing the project budget to accommodate a changed condition may be impractical. Third, placing the risk on the contractor provides the contractor with an incentive to minimize the financial effect of the discovered condition. If the contract has a differing site conditions clause, the contractor may see the changed condition as an opportunity to recoup other losses on the project at the owner's expense. Finally, in a competitive market, empirical evidence indicates that contractors do not quantify the risk of differing site conditions and may undervalue the risk. Under these conditions, elimination of the differing site conditions clause benefits the owner at little or no cost.

Whether to include a differing site conditions clause must, therefore, be based upon a review of the project, the potential risks, the available resources, and the owner's ability to absorb the risks.

Federal differing site conditions clause. The federal differing site conditions clause is the model from which most differing site conditions clauses

---

[2] *See* G.E. Mason, Stanford University, A Quantitative Risk Management Approach to the Selection of Construction Contract Provisions, Technical Report No. 173 (1973).

spring. Federal contracts contain a standard provision relating to differing site conditions,[3] which takes precedence over any contrary language in the contract.[4] Clause a(1) (usually referred to as a Type I claim) provides for an equitable adjustment if the conditions encountered differ materially from those indicated in the contract. Although the representation of the conditions need not be explicit, the contract documents must provide sufficient grounds to justify a bidder's expectation of latent conditions materially different from those actually encountered.[5]

When the contract documents do not contain affirmative misrepresentations as to anticipated conditions, a contractor's right to a contract adjustment may nonetheless arise from unusual physical conditions differing materially from those ordinarily encountered in work of the character provided in the contract.[6] Such claims, generally referred to as Type II claims, do not usually permit a contract adjustment for unanticipated weather conditions.[7]

AIA/EJCDC differing site conditions clauses. The 1987 edition of the American Institute of Architects (AIA) General Conditions for the Contract for Construction contains a differing site conditions clause similar to the federal model.[8] The Engineers Joint Contract Documents Committee (EJCDC)

---

[3] 48 C.F.R. § 52.236-2 (1991).

(a) The Contractor shall promptly, and before such conditions are disturbed, give a written notice to the Contracting Officer of: (1) subsurface or latent physical conditions at the site which differ materially from those indicated in this contract, or (2) unknown physical conditions at the site of an unusual nature, which differ materially from those ordinarily encountered and generally recognized as inhering in work of the character provided for in this contract.

(b) The Contracting Officer shall investigate the site conditions promptly after receiving the written notice. If the conditions do materially so differ and cause an increase or decrease in the Contractor's cost of, or of the time required for, performing any part of the work under this contract, whether or not changed as a result of the conditions, an equitable adjustment shall be made under this clause and the contract modified in writing accordingly.

(c) No request by the Contractor for an equitable adjustment to the contract under this clause shall be allowed unless the Contractor has given the written notice required; provided, however, the time prescribed in (a) above for giving written notice may be extended by the Contracting Officer.

(d) No request by the Contractor for an equitable adjustment to the contract for differing site conditions shall be allowed if made after final payment under this contract.

[4] Fattore Co. v. Metropolitan Sewerage Comm'n, 505 F.2d 1 (7th Cir. 1974).

[5] Foster Constr. C.A. v. United States, 435 F.2d 873 (Ct. Cl. 1970); P.J. Maffei Bldg. Wrecking Corp. v. United States, 732 F.2d 913 (Fed. Cir. 1984).

[6] 48 C.F.R. § 52.236.2 (1991).

[7] Turnkey Enters., Inc. v. United States, 597 F.2d 750 (Ct. Cl. 1979); Banks Constr. Co. v. United States, 364 F.2d 357 (Ct. Cl. 1966).

[8] AIA A201-1987, ¶ 4.3.6, Claims for Concealed or Unknown Conditions.

Standard General Conditions of the Construction Contract[9] contains a differing site conditions clause that departs from the federal model in several significant ways. First, "technical data" such as soils reports are identified in the Special Conditions but are not contract documents.[10] The contractor may rely upon facts, such as boring logs, being accurate but may not rely upon any of the opinions, conclusions, or interpretations contained in the "technical data."[11] This is intended to preclude the contractor's argument that the owner breached its agreement by providing misleading reports. If the facts are materially inaccurate, however, a claim can be made for a differing site condition.[12] In contrast, the federal differing site conditions clause does not discuss noncontract document references. A contractor claiming error in technical data would have to rely upon theories of misrepresentation, rather than claiming directly under the federal differing site conditions clause.

Second, the EJCDC differing site conditions clause permits a claim when the subsurface or physical condition revealed "is of such a nature as to require a change in the Contract Documents."[13] *Contract documents* is defined very broadly in the EJCDC General Conditions and includes the contractor's bid, as well as the contract and the plans and specifications.[14] It will be relatively easy for a contractor to assert that a discovered condition requires a change in the plans or specifications. Under a strict reading of this provision, a contractor could even assert that a differing site condition existed because the discovered condition required a change in its bid. Owners will wish to delete this provision.

Third, the EJCDC differing site conditions clause permits a recovery when the discovered conditions differ materially from those "shown or indicated" in the contract documents.[15] In contrast, the federal differing site conditions clause only permits equitable adjustment when the discovered conditions differ materially from those "indicated" in the contract documents.

The scope of a differing site conditions clause can be narrowed by site inspection clauses and nonreliance clauses as discussed below.

Site Investigation Clauses

Courts have found an implied obligation that a contractor make at least a minimal inspection of the site to familiarize itself with the property.[16] However, most modern contracts include an express "site inspection clause" obligating the

---

[9] EJCDC 1910-8 (1990).

[10] *Id.* ¶ 4.2.2.

[11] *Id.*

[12] *Id.* ¶ 4.2.3.1.

[13] *Id.* ¶ 4.2.3.2.

[14] *Id.* ¶ 1.10.

[15] EJCDC 1910-8 (1990), ¶ 4.2.3.3.

[16] Mojave Enters. v. United States, 3 Cl. Ct. 353 (1983).

contractor to inspect and familiarize itself with the conditions at the site. The AIA[17] and EJCDC[18] contract documents both contain such site inspection clauses, although their approach differs. AIA Document A201 directs the contractor to verify field conditions and measurements before commencing construction.[19] In contrast, EJCDC 1910-8 precludes contractor claims if the differing site condition could reasonably have been discovered by the contractor while performing required examinations, investigations, explorations, or tests.[20]

Application of the differing site conditions clause must consider the requirements of the site inspection clause. When a site inspection clause exists and the contractor unreasonably fails to avail itself of the site inspection opportunity, the contractor may be foreclosed from invoking the terms of the differing site condition provision.[21] If, however, the contractor makes a reasonable inspection of the site yet fails to discover the differing site condition, the two clauses apparently conflict.

Courts have traditionally resolved the conflict by applying a standard of reasonableness: the contractor is obligated only to discover conditions apparent through a reasonable investigation.[22] The contractor is not obligated to discover hidden conditions not amenable to reasonable investigation, nor is it obligated to perform burdensome, extensive, or detailed tests or analyses.[23] The obligation to investigate the site may be further limited by the surrounding circumstances. If the investigation is constrained by weather conditions, site conditions, or time restraints in the contracting process, the contractor will only be required to perform such an investigation as is reasonable under the circumstances.[24]

Site inspection clauses do not require the contractor to rely only on information gained through the site inspection and to disregard all information received from the project owner. Unless barred by the terms of the contract,[25] the contractor is entitled to rely on information received from the project owner, particularly when the contractor does not have adequate time or opportunity to

---

[17] AIA A201-1987, ¶¶ 3.2.1, 3.2.2.

[18] EJCDC 1910-8 (1990), ¶ 4.2.6.4.

[19] AIA A201-1987, ¶ 3.2.2.

[20] EJCDC 1910-8 (1990), ¶ 4.2.6.4.2.

[21] Clark v. United States, 5 Cl. Ct. 447 (1984), *aff'd,* 770 F.2d 180 (Fed. Cir. 1985).

[22] Foster Constr. C.A. v. United States, 435 F.2d 873, 888 (Ct. Cl. 1970); Farnsworth & Chambers Co. v. United States, 346 F.2d 577, 580–81 (Ct. Cl. 1965).

[23] Foster Constr. C.A. v. United States, 435 F.2d 873, 888 (Ct. Cl. 1970); Farnsworth & Chambers Co. v. United States, 346 F.2d 577, 580–81 (Ct. Cl. 1965), providing:

> The contractual requirement that [the contractor] make its own investigation of the site does not obliterate the Changed Conditions clause, nor did this requirement obligate [bidding contractors] to discover, at their peril, subsurface conditions hidden by the river's water and thus unavailable to any reasonable preaward inspection.

[24] Frederick Snare Corp. v. Maine-New Hampshire Interstate Bridge Auth., 41 F. Supp. 638 (D.N.H. 1941).

[25] See following discussion of reliance clauses.

conduct a thorough independent investigation or when requiring an extensive investigation would not be reasonable under the circumstances.

Finally, although the majority of courts limit site inspection clauses by a standard of reasonableness, some commentators suggest the limitation should be made explicit. Contractors may wish to supplement site inspection clauses with language limiting the obligation to expressly exclude extensive or detailed testing.[26]

Disclaimer or Reliance Clauses

Disclaimer or reliance clauses may also limit the effectiveness of a differing site conditions clause. Such clauses typically state that information received from the project owner is provided solely for informational purposes and that the owner does not warrant the accuracy or sufficiency of the information provided. The objective of such provisions is to render unreasonable any reliance by the contractor on the information in characterizing the condition of the property.

Courts have reached a variety of results on the effect of disclaimer provisions. At least two cases have held a disclaimer was effective to preclude a contractor from arguing that reliance on the information received from the project owner was reasonable.[27] In order to be effective, such clauses should provide that the information was not warranted and that the contractor has not relied on the information. Such provisions are most effective when combined with a site inspection clause.

However, most courts hold that disclaimer clauses do not preclude reliance on information received from the project owner. The situations in which courts have allowed contractors to rely on information received from the project owner despite a disclaimer clause may be grouped into three categories. First are those cases holding that reliance was permissible because the contractor performed a reasonable investigation that confirmed or supported the information received from the project owner.[28] Second are those cases holding that

---

[26] O'Leary, *Negotiating the General Construction Contract: Supplemental General Conditions, in* Design and Construction Contracts: New Forms—New Realities, 1987. A.B.A. Sec. Real Property, Probate & Trust Law 249, suggests including the following language:

> But such general familiarity shall not be construed to be a representation that the Contractor is familiar with conditions which would normally require extensive or detailed testing in order for discovery of such conditions to occur, such as, but not limited to, subsoil or subsurface conditions not apparent without detailed soil boring, engineering, or other tests.

[27] J.E. Brenneman Co. v. Commonwealth Dep't of Transp., 56 Pa. 210, 424 A.2d 592 (1981); Zurn Eng'rs v. State, 69 Cal. App. 3d 798, 138 Cal. Rptr. 478, *cert. denied,* 434 U.S. 985 (1977).

[28] Morrison-Knudsen Co. v. United States, 397 F.2d 826 (Ct. Cl. 1968); J.F. Shea Co. v. United States, 4 Cl. Ct. 46, 52 (1983), *aff'd,* 754 F.2d 338 (Fed. Cir. 1985).

reliance was justified because the project owner intended that the contractor rely on the information in preparing a bid.[29] Third are those cases holding that reliance was justified because the circumstances did not allow sufficient time for the contractor to conduct an adequate independent investigation.[30] The cumulative effect of these limitations is that a contractor may rely on information received from the project owner except when relatively simple inquiries might have revealed contrary conditions.[31]

However, even an otherwise effective disclaimer clause will not shelter a project owner from a claim for an affirmative misrepresentation.[32]

§ 1.5 —Limitations on Economic Consequences

Differing site conditions, site inspection, and reliance clauses focus on the liability of the parties for differing site conditions. Another approach is to assume that differing site conditions claims can and will be made and to limit the amount and type of damages that can be awarded. Unit price, per diem delay, and no-damages-for-delay clauses are examples of this approach.

Unit Price Contracts

When differing site conditions are a possibility, unit price bidding may be preferable to a fixed-price agreement. Under a unit price agreement, key elements, such as excavation, are priced by formula. The contractor is paid for the quantity installed or removed times the agreed unit price for that item. If quantities increase or decrease as a result of the differing site condition, the adjustment to the contract is relatively simple. By using unit prices, the owner obtains the benefit of a competitively bid price on the potential extra work.

Several issues should be considered in drafting unit price provisions. First, the contract should compute the contractor's compensation for given items on a unit price rather than a lump sum basis. It should also state that any increase or decrease in quantities is not a change to the contract. Second, the unit prices set forth in the agreement should state what markups, if any, are payable on the unit price work. Third, if significant variations in quantity are possible, unit prices

[29] Christie v. United States, 237 U.S. 234 (1915); Eastern Tunneling Corp. v. Southgate Sanitation Dist., 487 F. Supp. 109 (D. Colo. 1979).

[30] Christie v. United States, 237 U.S. 234 (1915); Eastern Tunneling Corp. v. Southgate Sanitation Dist., 487 F. Supp. 109 (D. Colo. 1979); Frederick Snare Corp. v. Maine-New Hampshire Interstate Bridge Auth., 41 F. Supp. 638 (D.N.H. 1941).

[31] Foster Constr. C.A. v. United States, 435 F.2d 873, 882 (Ct. Cl. 1970); Frederick Snare Corp. v. Maine-New Hampshire Interstate Bridge Auth., 41 F. Supp. 638 (D.N.H. 1941).

[32] Eastern Tunneling Corp. v. Southgate Sanitation Dist., 487 F. Supp. 109, 114–15 (D. Colo. 1979).

and overhead assessments may differ depending upon the final quantity. The owner may wish to set unit prices for different ranges of quantity. Fourth, a procedure should exist for auditing the actual quantities placed, installed, or removed. Often this will involve review by the designer or construction manager. It may also be prudent to specify the method of calculation. For instance, should excavation be measured by truck loads or by an engineer's survey and calculation of material moved?

British practice extends the unit price concept to its logical conclusion. The owner retains a professional quantity surveyor who prepares a Bill of Quantities for the project based upon the local Standard Form of Measurement. The contractor bids on the Bill of Quantities on a unit price basis rather than performing its own quantity takeoffs. If changes are made to the project due to differing site conditions, the additions or deletions to the contract are comparatively simple to perform. Architects practicing under both the British and American systems report that it is easier to resolve differing site conditions claims under the quantity survey approach than under a lump sum bid system.

Per Diem Delay

The unit price concept can be extended to delay damages as well. As a part of the bid, the contractor can be required to include a per day calculation of delay damages. This includes, but is not limited to, field overhead, home office overhead, idle labor and equipment, and any loss of efficiency or impact on later performance. The per day bid amount becomes a liquidated damage for any owner-caused delay. To determine the low bidder, an assumed number of delay days appropriate to the project, multiplied by the individual bidder's per diem delay price, is added to each bid. Per diem delay provisions are particularly appropriate when delay is anticipated due to uncertain site conditions and has been used successfully in major renovation projects.[33]

Per diem bidding of delay reduces the impact of a differing site condition claim by providing a competitive assessment of the contractor's overhead costs. Unlike a claim situation, the contractor, when bidding, has an incentive to include a low overhead calculation. Evaluation of overhead is often one of the most hotly disputed elements of delay claim litigation. The per diem approach also reduces the effect of impact damages by incorporating all aspects of the delay claim into a single daily charge.

No Damages for Delay

No-damages-for-delay clauses can be effective in limiting the impact of differing site conditions by transferring the economic risk of delay to the contractor.

---

[33] White, No Damage for Delay—An Appropriate Response for Delay Claims (1988) (unpublished paper, ABA Forum on the Construction Industry, Fourth Annual Meeting).

Theoretically, the contractor will knowingly charge a premium for accepting this risk while preparing its bid. This front-loaded cost may be more efficient because the contractor is capable of assessing the risks and, more important, capable of minimizing the impact of any delays that occur during construction.

In *Dobson v. State,*[34] the trial court granted the contractor delay damages due to 15 uncharted underground obstacles. On appeal, the court reconciled the no-damages-for-delay clause with the contract's differing site conditions clause, stating:

> The judge found that on the basis of the contract language and surrounding circumstances, the parties did not intend the no delay damages provision to apply to delay damages from uncharted underground obstacles. He reasoned paragraph 15B.27, providing that costs arising from uncharted obstacles could be adjusted as they arose, negated any intent to include them in the coverage of paragraph 12E. We disagree with this interpretation of the contract.
>
> The reasonable and sensible interpretation is that the parties intended that there would be no delay damages even where uncharted obstacles were discovered. When such objects were discovered, the parties intended for the contractor to submit a change order request and the parties would then agree upon the "mutually acceptable additional compensation," if necessary, caused by the uncharted obstacles. Plaintiff submitted change orders due to the uncharted obstacles and was paid for all of the additional actual expenses incurred. Plaintiff was not paid, however, for delay damages.

Thus, the no-damages-for-delay clause effectively limits the cost of a differing site conditions claim by eliminating recovery for the associated project delays. Because of the potentially draconian effect of a no-damages-for-delay clause, contractors have aggressively contested their application and have sought legislation limiting their use. The remainder of this section examines the enforceability of no-damages-for-delay clauses together with the principal exceptions to their application.

Enforceability of no-damages-for-delay clauses. The general rule is that no-damages-for-delay clauses are enforceable, subject to the exceptions to be discussed.[35] The basis for this rule is threefold. First, the parties, after conducting arm's-length negotiations, should be free to allocate risks between them.[36] The second commonly stated purpose is the avoidance of "'vexatious' litigation as to whether delays are reasonable or unreasonable or, for that matter, real or

---

[34] 218 N.J. Super. 123, 526 A.2d 1150 (1987).

[35] *See* Brunner, Annotation, *Validity and Construction of "No Damage" Clause with Respect to Delay in Building or Construction Contract,* 74 A.L.R. 3d 187 (1976).

[36] One court has stated that enforcing the no-damages-for-delay clause is a "frank recognition of such a perfectly common and acceptable business practice, by which an entrepreneur may provide protection against its own fault." Kalisch-Jarcho, Inc. v. City of N.Y., 58 N.Y.2d 377, 384, 448 N.E.2d 413, 416, 461 N.Y.S.2d 746, 749 (1983).

fancied."[37] The final, and perhaps most convincing, justification is that the clause will discourage "dilatoriness itself" by the contractor.[38]

Courts have broadly defined *delay damages* when interpreting no-damages-for-delay clauses in contracts.[39] Contractors have thus been frustrated in their attempts to characterize their losses as falling outside the scope of the delay damages clause.[40] Typically, a contractor will allege that its damages arise from an owner's breach of contract and are not merely delay damages.

Corinno Civetta Construction Corp. v. City of New York (Honeywell)[41] provides an example of this attempted maneuver. Honeywell contended that the city of New York had breached its contract by failing to correct incompatibilities between Honeywell's equipment and equipment to which it connected and in failing to prepare plans showing conduit necessary for Honeywell's work. Rather than treat Honeywell's claim as independent of delay damages, the New York Court of Appeals characterized breach of contract as a narrow exception to the enforceability of no-damages-for-delay clauses, stating:

> A contractee's breach of contract is also recognized as an exception to the enforceability of exculpatory clauses but, as with abandonment and for the same reason, the exception is applied to an especially narrow range of circumstances. Because the exculpatory clause is specifically designed to protect the contractee from claims for delay damages resulting from its failure of performance in ordinary, garden variety ways, delay damages may be recovered in a breach of contract action only for the breach of a fundamental, affirmative obligation the agreement expressly imposes on the contractee.[42]

Having established this stern standard, the court went on to hold that Honeywell had not proven breach of a fundamental, affirmative obligation.

Exceptions to no-damages-for-delay clauses. Because courts have broadly construed the scope of no-damages-for-delay clauses, the exceptions to their enforceability are significant. In general, the no-damages-for-delay clause will not be enforced if the delays that occurred were not contemplated when the contract was executed.[43] Furthermore, the contractor's delay claim will not be

---

[37] *Id.*, 448 N.E.2d at 416, 461 N.Y.S.2d at 749.

[38] *Id.*

[39] *See* Western Eng'rs, Inc. v. State Road Comm'n, 20 Utah 2d 294, 437 P.2d 216 (1968); Martin Mechanical v. Carlin, 132 A.D.2d 688, 518 N.Y.S.2d 166 (1987).

[40] *See, e.g.*, Bates & Rogers Constr. Co. v. Greeley & Hansen, 109 Ill. 2d 225, 230, 486 N.E.2d 902, 905 (1985).

[41] 67 N.Y.2d 297, 493 N.E.2d 905, 502 N.Y.S.2d 681 (1986).

[42] *Id.*, 493 N.E.2d at 912, 502 N.Y.S.2d at 688.

[43] *Cf.* Carrabine Constr. Co. v. Chrysler Realty Corp., 25 Ohio St. 3d 222, 495 N.E.2d 952 (1986). (It was not an uncontemplated delay when construction was delayed seven months because of an inability to obtain the necessary building permits.)

barred if the delays were caused by the contractee's active interference, bad faith, or intentional misconduct.[44] An abandonment of the contract will also expose the owner and design professional to delay damages, regardless of the no-damages-for-delay clause.[45] Finally, a material misrepresentation or omission will support a delay claim despite a contractual limitation on damage claims.

Active interference, bad faith, or intentional misconduct. The owner's motives may be the key to establishing this exception. Although the term *active interference* seemingly would not require intentional misconduct, recent cases have imposed a scienter requirement.

In *Dobson v. State,*[46] a contractor was required to install low-pressure steam distribution lines manufactured by a specific manufacturer or an approved substitute. When orders for the material were placed, the manufacturer placed intolerable terms and conditions on the order. The contractor requested a substitution, but the state's contracting officer refused approval of the change, threatening to purchase the material, back charge the contractor, and impose penalties. Under compulsion, the contractor acceded to the manufacturer's demands. The project was delayed when, among other things, the manufacturer did not timely deliver the pipe.

The contractor argued, and the trial court found, that the contracting officer's actions amounted to active interference and bad faith. Although agreeing that active interference was an exception to enforcement of a no-damages-for-delay clause, the superior court reversed, stating:

> "Active interference" connotes more than negligence. While the term is not capable of precise definition, it contemplates reprehensible behavior beyond "a simple mistake, error in judgment, lack of total effort, or lack of complete diligence . . . ." *Peter Kiewit Sons' Co. v. Iowa Southern Utilities Co.,* 355 F. Supp. 376, 399 (S.D. Iowa 1973). The public agency must commit some affirmative, willful act, in bad faith which unreasonably interferes with the contractor's compliance with the contract terms before it can be said that there has been active interference which subjects the public agency to delay damages notwithstanding the no damage for delay clause in the contract. *Ibid.* The ultimate determination must be based on the intention of the parties, *Buckley & Co., Inc. v. State,* 140 N.J. Super. 289, 299 (Law Div. 1975), as can be discerned from the contractual language in light of the circumstances.[47]

---

[44] *Cf.* Gust K. Newberg, Inc. v. Illinois State Toll Highway Auth., 153 Ill. App. 3d 918, 506 N.E.2d 658 (1987) (No bad-faith conduct because record showed that owner proceeded "as expeditiously as it could" to try to solve the delays.)

[45] *See, e.g.,* Corinno Civetta Constr. Corp. v. City of New York, 67 N.Y.2d 297, 493 N.E.2d 905, 502 N.Y.S.2d 681 (1986).

[46] 218 N.J. Super. 123, 526 A.2d 1150 (1987).

[47] *Id.,* 526 A.2d at 1153.

In *Kalish-Jarcho v. City of New York,*[48] the New York Court of Appeals similarly held that "active interference" requires an element of wrongdoing. The contractor in *Kalish-Jarcho* agreed to install a heating, ventilating, and air conditioning (HVAC) system for a police headquarters in New York City. Its contract was for approximately $8 million and provided that construction was to be completed within 36 months. However, because of "'endless' revisions of scores of plans and drawings . . . [and] other acts of omission or commission interfering with the sequence and time of the work," plaintiff was substantially delayed. The project was extended an additional 28 months, and plaintiff sought damages of over $3.3 million, or approximately 40 percent of the original contract amount.

The trial court instructed the jury that plaintiff could not recover its delay damages unless it could prove active interference by the owner. The basis of this ruling was the no-damages-for-delay clause in the contract, which read as follows:

The Contractor agrees to make no claim for damages for delay in the performance of this contract occasioned by any act or omission to act of the City or any of its representatives, and agrees that any such claim shall be fully compensated for by an extension of time to complete performance of the work as provided herein.

The jury awarded plaintiff $806,382 based on its finding that the owner had actively interfered with the contractor's work.

The New York Court of Appeals reversed, holding that the no-damages-for-delay clause was enforceable provided that the owner's conduct did not "smack of intentional wrongdoing."[49] The trial court's standard of "active interference" was rejected because it did not suggest wrongdoing or "connote willfulness, maliciousness, abandonment, [or] bad faith" by the owner.[50] The basis for the court's ruling was its belief that the owner and contractor, while executing their contract, intended that the clause be applicable when the owner actively interfered with the contractor's work.[51]

Dobson and *Kalish-Jarcho* demonstrate the difficulty a contractor may confront when attempting to argue that delay was caused by the owner or its representatives.[52] Without proof of actual malice or at least a conscious disregard of

---

[48] 58 N.Y.2d 377, 448 N.E.2d 413, 461 N.Y.S.2d 746 (1983).

[49] *Id.* at 385, 448 N.E.2d at 416, 417, 461 N.Y.S.2d at 750.

[50] *Id.* at 386, 448 N.E.2d at 417, 461 N.Y.S.2d at 750, 751.

[51] The jury also found that the delay that occurred was within the contemplation of the parties at the time the contract was made. Interestingly, the jury further determined that the delay was caused by both parties, placing 63% of the blame on the city and 37% on the contractor. *Id.* at 382, 448 N.E.2d at 415, 461 N.Y.S.2d at 748.

[52] Because the owner's motive is an issue in an "active interference" case, the contractor can use extrinsic evidence of similar practices on other projects or abusive payment practices. Newberry Square Dev. Corp. v. Southern Landmark, Inc., 578 So. 2d 750, 752–53 (Fla. Dist. Ct. App. 1991).

the contractor's rights, there is a strong likelihood that the no-damages-for-delay clause will be applied.

Unreasonable or uncontemplated delays. A no-damages-for-delay clause has allowed an owner to successfully defend a contractor's delay claim even though the owner caused an eight-year delay to a scheduled two-year project.

In *Buckley & Co. v. City of New York*[53] the plaintiff contractor was awarded a contract to construct a pumping station for $8,442,130. Construction was scheduled to begin by June 1966 and was to be completed in June 1968. Immediately after construction commenced, the project was delayed because the owner's design of a cofferdam improperly allowed water to seep into the excavation. The delay was substantial: the project was not completed until 1976, eight years after the original scheduled completion date.[54]

Stating that the no-damages-for-delay clause reflected "the parties' unambiguous intent to have the contractor absorb the costs of the contractee's delay," the court barred plaintiff from recovery of its delay damages.[55] The delay was not uncontemplated because the contract provided that the contract was to be modified if unforeseen or unanticipated subsurface conditions occurred.[56] Moreover, the court found that the owner was not guilty of willful misconduct, stating that the owner "was at most simply negligent in its evaluation of the subsurface conditions and design of the cofferdam. Delays caused by the contractee's ordinary negligence, however, fall well within the scope of the [no-damages-for-delay clause] and may not, therefore, support a claim for delay damages."[57]

In determining whether a particular delay was contemplated, courts often look to the language of the contract itself. Thus, in *Dobson v. State of New Jersey*,[58] the court held that a clause that advised the contractor that existing data concerning the location of utilities might be inaccurate sufficiently alerted the contractor to the possibility of uncharted obstacles. Similarly, in *Gust K. Newberg, Inc. v. Illinois State Toll Highway Authority*,[59] the court relied upon a provision requiring the contractor to acquaint itself with the status of right-of-way acquisition before submitting a bid. In these instances, it appears that the

---

[53] 121 A.D.2d 933, 505 N.Y.S.2d 140 (1986).

[54] The opinion does not state the amount of damages sought by the contractor. However, inflation alone would more than double the contract price.

[55] Buckley & Co. v. City of N.Y., 121 A.D.2d 933, 505 N.Y.S.2d 140, 142 (1986).

[56] The opinion does not state the exact terms of the "unforeseen condition" clause. Normally, such clauses provide that the contractor is entitled to its extra direct costs incurred as a result of the unexpected condition.

[57] Buckley & Co. v. City of N.Y., 121 A.D.2d 933, 936, 505 N.Y.S.2d 140, 143 (1986).

[58] 218 N.J. Super. 123, 526 A.2d 1150 (1987).

[59] 153 Ill. App. 3d 918, 506 N.E.2d 658 (1987).

warnings were specifically related to the project in question and were conspicuous. As such, the contractor could evaluate the actual risk.

In *Corinno Civetta Construction Corp. v. City of New York*[60] however, the court relied upon a boilerplate provision to refute the contractor's argument that the delays caused by interferences were not contemplated:

> Although plaintiff suggests that this problem was not contemplated by the parties and therefore that it need not prove gross negligence or abandonment to avoid the exculpatory clause, the contract unambiguously proves otherwise. Indicative is a clause in the General Conditions which specifies that all contractors will exchange working drawings, examine them and report "any interferences or objections" to the project engineer "in order to avoid delays in [the] progress of construction."[61]

Whether delay from a particular problem was contemplated should not be dependent on boilerplate provisions. General conditions are not usually subject to negotiation between owner and contractor. The contractor first sees the general conditions as part of a large bid package that must be quickly reviewed and priced. There is no practical opportunity to review all of the conditions and determine the specific applicability of each to the project. Even if there were, bids that take exception to specific contract language may be rejected as nonconforming. Thus, the contractor has little opportunity or incentive to object to the exculpatory language.

The contractor would not, in many instances, know which clauses are important to a specific project. General conditions are usually drafted as boilerplate forms that are overly broad and only minimally tailored to the project. The similarity of general conditions from project to project belies the contention that a specific limitation gives ample notice of a potential problem on a specific project.[62] One can only hope that courts will recognize the distinction between specific admonitions and boilerplate requirements.

Abandonment of contract. Two types of abandonment are relevant to delay damage claims. In the first, the owner so delays completion as to constructively abandon completion of the project. In the second, excessive changes to the work, either in number or scope, so radically change the nature of the contract that it would be unjust to hold the parties to their original agreement. In the latter

---

[60] 67 N.Y.2d 297, 493 N.E.2d 905, 502 N.Y.S.2d 681 (1986).

[61] *Id.*, 493 N.E.2d at 915, 502 N.Y.S.2d at 691.

[62] As an example, AIA A201-1987, General Conditions of the Contract for Construction, requires the contractor to review the designer's plans for inconsistencies (paragraph 3.2.1) and to take field measurements and compare them to the contract documents (paragraph 3.2.2). Should these requirements mean that delays from such inconsistencies are contemplated by the parties? The author believes that this type of clause is too general to be probative of the parties' intent with regard to waiver of delay damages.

situation, it is said that the contract has been abandoned although the work has not. Because there is no contract between the parties, the contractor may perform free of limitations and seek recovery on a quantum meruit basis.

Abandonment of the project can be difficult to prove. In *Corinno Civetta,* the court recognized that a no-damages-for-delay clause would not be given effect when the owner causes delays "which are 'so great or so unreasonable that they may fairly be deemed equivalent to [its] abandonment of the contract.'"[63] The court strictly construed this exception, however, by stating:

> Thus, to avoid the risk of the exculpatory clause and recover on the ground of abandonment, a contractor must establish that the contractee is responsible for delays which are so unreasonable that they connote a relinquishment of the contract by the contractee with the intention of never resuming it.[64]

The court reviewed the evidence surrounding Honeywell's claim and concluded that the city was actively pursuing the project.

> The city's activities in attempting to complete this project as soon as possible despite numerous problems and delays and its acceptance of plaintiff's work at three out of the five sites of this interrelated project refutes any claim of abandonment based upon the panel boards.[65]

It is interesting to note that "reasonableness" of the delays, which is part of the initial formulation of the abandonment exception, was omitted from the court's review of the evidence. Rather, the focus shifted to the intention to proceed with the project. Because the public agency will almost always contend that it still desires completion, it is difficult to visualize a situation in which a New York court would find abandonment.

Misrepresentations. When a contractor can show justifiable reliance upon the owner's statements or representations, it may be possible to obtain delay damages in the face of a no-damages-for-delay clause.

In *P.T.&L. Construction Co. v. State,*[66] a joint venture entered into an agreement to construct a 1.4-mile section of Interstate 78. When construction commenced, the contractor discovered that the site drained poorly and that the west end of the site was sometimes under three to four feet of water. As a result, stripping operations, which were to be completed in 3 days, actually required 171 days. The contractor sought recovery for the additional delay and for delays resulting from poor coordination of site utilities.

[63] Corinno Civetta Constr. Corp. v. City of New York, 67 N.Y.2d 297, 493 N.E.2d 905, 912, 502 N.Y.S.2d 681, 688 (1986).

[64] *Id.* (emphasis added).

[65] *Id.,* 493 N.E.2d at 915, 502 N.Y.S.2d at 691.

[66] 108 N.J. 539, 531 A.2d 1330 (1987).

The contract documents required a type of fill material used only in dry conditions. In addition, there were indications in the documents that drainage on the site was adequate. There was no indication in the contract documents that excessively wet conditions existed. During discovery, it became known that before design of the project, the state had received a letter stating that "the extent of the removal of the wet excavation . . . will depend on the climatic conditions and the time of the year in which the construction will be accomplished."[67] In addition, the agency was aware that the design was predicated upon completion of a project to lower a branch of an adjacent river. The contractor argued that the site conditions were misrepresented by the contract documents and that the state had failed to disclose information relevant to construction of the project.

In a lengthy opinion, the court carefully distinguished between a "positive averment describing the actual condition of the land, . . . or a more general description of site conditions, qualified by a disclaimer."[68] Applied to the facts of the case, the court found that the implications that dry conditions prevailed were not "positive averments" and thus would not circumvent the clause-limiting claims arising from subsurface conditions. The letter and the knowledge of the companion river project, however, raised the issue of a constructive misrepresentation.

The court noted that under certain circumstances, a governmental agency could be liable for failing to impart its knowledge of the difficulties to be encountered in the construction project. Although there was no indication that the information had been deliberately withheld, the court affirmed the trial court's determination that the omissions were material misrepresentations and allowed recovery of extra costs due to the extra excavation efforts. Thus, clauses limiting recovery for subsurface conditions were not applicable.

In jurisdictions where public agencies have a duty to disclose material information,[69] a contractor can attempt to avoid the no-damages-for-delay limitation by arguing an actual or constructive misrepresentation. An owner wishing to oppose this contention should provide contractors with the opportunity, before bid, to review the files maintained by the agency and its designers. In this way, the owner can argue that the contractor did not reasonably rely upon the failure to provide material information.

Because of the harsh results rendered by the courts, and also undoubtedly because of effective lobbying by contractors, the legislatures of Washington, California, and Oregon have enacted statutes limiting the owners' right to require the contractor to waive its delay damage claim.

---

[67] *Id.* at 559, 531 A.2d at 1340.

[68] *Id.* at 539, 531 A.2d at 1339.

[69] *See, e.g.,* Warner Constr. Corp. v. City of Los Angeles, 2 Cal. 3d 285, 293, 466 P.2d 996, 1001, 85 Cal. Rptr. 444, 449 (1970); Samo, Annotation, *Public Contracts: Duty of Public Authority to Disclose to Contractor Information, Allegedly in its Possession, Affecting Cost or Feasibility of Project,* 86 A.L.R. 3d 182 (1978).

In *Christiansen Brothers, Inc. v. State,*[70] the Washington Supreme Court affirmed the trial court's bar of the contractor's delay claim based on the no-damages-for-delay clause in the contractor's contract with the owner. The delay, caused by architectural design errors, among other things, was for approximately 10 months, and the contractor's total claim was for $227,753, approximately 7 percent of the original contract amount. The supreme court dismissed the contractor's argument that the delay clause was unconscionable because it was undisputed that the contractor had notice before submission of its bid and could adjust the bid accordingly.

Immediately after *Christiansen* was decided, the Washington legislature enacted a statute, which, on its face, voids no-damages-for-delay clauses for "unreasonable delays" in all construction projects:

> Any clause in a construction contract . . . which purports to waive, release, or extinguish the rights of a contractor, subcontractor, or supplier to damages or an equitable adjustment arising out of unreasonable delay in performance which delay is caused by the acts or omissions of the contractee or persons acting for the contractee is against public policy and is void and unenforceable.
>
> This section shall not be construed to void any provision in a construction contact . . . which (1) requires notice of delays, (2) provides for arbitration or other procedure for settlement, or (3) provides for reasonable liquidated damages.[71]

The California and Oregon legislatures have adopted narrower statutes, which only apply to public contracts. The Oregon statute provides:

> (1) Any clause in a public contract for a public improvement that purports to waive, release or extinguish the rights of a contractor to damages or an equitable adjustment arising out of unreasonable delay in performing the contract, if the delay is caused by acts or omissions of the public contracting agency or persons acting therefor, is against public policy and is void and unenforceable.
>
> (2) Subsection (1) of this section is not intended to render void any contract provision that:
>
> (a) Requires notice of any delay;
>
> (b) Provides for arbitration or other procedures for settlement of contract disputes; or
>
> (c) Provides for reasonable liquidated damages.[72]

In 1987, California amended its statute to prohibit waiver of the statute's provisions. As amended, the statute states:

---

[70] 90 Wash. 2d 872, 586 P.2d 840 (1978).

[71] Wash. Rev. Code § 4.24.360 (1988 & Supp. 1992).

[72] Oregon Rev. Stat. § 279.063 (1989).

> Contract provisions in construction contracts of public agencies and subcontracts thereunder which limit the contractee's liability to an extension of time for delay for which the contractee is responsible and which delay is unreasonable under the circumstances involved, and not within the contemplation of the parties, shall not be construed to preclude the recovery of damages by the contractor or subcontractor.
>
> No public agency may require the waiver, alteration, or limitation of the applicability of this section. Any such waiver, alteration, or limitation is void. This section shall not be construed to void any provision in a construction contract which requires notice of delays, provides for arbitration or other procedure for settlement, or provides for liquidated damages.[73]

One can anticipate that contractors' organizations will increasingly attempt to obtain legislative relief from the impact of no-damages-for-delay clauses. They can expect stiff opposition in jurisdictions where no-damages-for-delay clauses are quite effective. In New York, for example, an effort was made to have a similar statute enacted, but it was defeated due to opposition from the city of New York.[74]

Termination for Convenience

Termination for convenience is the ultimate contract safety valve and has several uses relevant to differing site conditions claims. First, when it becomes apparent that the differing site condition makes it impossible to deliver the project within the owner's economic abilities or projections, the owner can abandon the project without further liability. Second, when the differing site condition needs to be studied before any determination about how or whether to proceed can be made, it may be less expensive to terminate the contractor and later contract for completion than to incur a delay claim from the existing contractor. Discovery of hazardous materials, for instance, may call for termination during assessment. Third, discovery of a difficult site condition may cause the owner to seek a contractor with greater expertise. Termination for convenience permits such a substitution.

A termination provision should contain a formula for compensating the contractor on termination for convenience. The provision may be based on percentage of completion including profit or may be based upon the reasonable value of labor and materials installed before termination. Generally, a terminated contractor is not compensated for post-termination unabsorbed overhead,[75] but may recover when the suspension preceding termination results in overhead being unabsorbed.[76] The termination provisions should, therefore, consider the

[73] Cal. Pub. Cont. Code § 7102 (West 1985 & Supp. 1992).

[74] Walker, *Statutory Response to "No Damage for Delay" Clauses,* 6 Constr. Law. (1986).

[75] A.C.E.S., Inc., ASBCA No. 21,417, 79-1 B.C.A. (CCH) ¶ 13,809 (1979).

[76] Southwestern Eng'g v. Cajun Elec. Power, 915 F.2d 972, 977 (5th Cir. 1990).

overhead recoverable on termination or suspension. Finally, the termination provision should explicitly state that the contractor waives any damages or claims other than as provided under the termination payment.

§ 1.6 —Procedural Provisions

The effect of a differing site conditions claim is also affected by procedural provisions of the contract. Notice provisions prevent the presentation of stale claims and allow the owner to investigate the cause and consequence of the differing site condition when information is fresh and before the costs are incurred. Failure to abide by the notice provisions may result in a loss of the differing site conditions claim. Dispute resolution clauses affect the cost of determining the differing sites conditions claim as they impose dispute resolution procedures less expensive than litigation. In addition, by determining disputes while the project is ongoing, the impact of the differing site conditions claim is often minimized. The parties' incentive to compromise is also greatest while the project is underway.

Notice Provisions

Most differing site conditions clauses require a contractor to give the project owner prompt notice of a differing condition, and some require written notice within a defined period. Typically, the contract also provides that failure to provide notice within the specified period will result in a denial of any claim for additional compensation.

Courts view such clauses with disdain and will enforce them only if the project owner can show actual harm or prejudice as a result of the delay. However, project owners can often establish injury or prejudice as a result of a lack of notice of a differing site condition when they did not have an equal opportunity to inspect. In one recent case, the Armed Services Board of Contract Appeals denied a contractor's claim for a differing site condition because the delayed notice denied the government the opportunity to verify the differing site condition.[77]

Courts have construed written notice requirements more strictly. Contractors may be precluded from recovering compensation for differing conditions if they fail to comply with the written notice or recordkeeping procedures set forth in the contract.[78] However, when the project owner receives actual notice of the differing condition, several courts have held the contractor's failure to provide

---

[77] Meisel Rohrbau, ASBCA No. 35,566, ______ B.C.A. (CCH) ¶ ______ (1991).

[78] Skopek Bros., Inc. v. Webster Hous. Auth., 11 Mass. App. 947, 416 N.E.2d 1006 (1981); Blankenship Constr. Co. v. North Carolina State Highway Comm'n, 28 N.C. App. 593, 222 S.E.2d 452, *cert. denied,* 290 N.C. 550, 230 S.E.2d 765 (1976).

procedurally correct or timely notice did not bar a claim under the differing site condition clause.[79]

Alternative Dispute Resolution Provisions

Alternative dispute resolution can significantly lessen the consequence of differing site conditions by greatly reducing the resolution cost and by permitting an early resolution of the claim. To achieve these goals, the contract provisions must require the use of alternative dispute resolution instead of litigation, as in the case of binding arbitration, or as a condition precedent to litigation, as in the case of mediation.

The dispute resolution provisions should also require early presentation of the claim. Stringent notice requirements should require the contractor to immediately advise the owner of the existence of a differing site condition. This initial notice should commence dispute resolution. Within a given period of time after the initial notice, the contractor should be required to propose an amendment to the contract setting forth any increase in price or time. If the owner does not concur with the proposed amendment, the owner may invoke the assistance of a third-party arbitrator or mediator. Which dispute resolution device is then used will depend upon the type of project and the parties' preferences. The advantages and disadvantages of the different dispute resolution procedures are discussed below.

Arbitration. Arbitration is the most common form of alternative dispute resolution in the construction industry. EJCDC and AIA standard form agreements, for example, require arbitration in accordance with the Construction Industry Rules of the American Arbitration Association (AAA).[80]

Arbitration under the Construction Industry Rules is not the best tool for reducing the impact of a differing site conditions claim because it is designed with a postproject hearing in mind. The procedures may be modified, however, and AAA is willing to administer any arbitration agreement according to its terms. To accomplish the goals of efficient, prompt resolution, a number of points should be considered for inclusion in a modified arbitration agreement:

1. Establish a specific timetable.
2. Specifically encourage single-issue arbitration. This lessens the need to wait until the end of a project to determine all of the issues that will be encompassed by the arbitration.

[79] Metropolitan Sewerage Comm'n v. R.W. Constr., Inc., 72 Wis. 2d 365, 241 N.W.2d 371 (1976); Frederick Snare Corp. v. Maine-New Hampshire Bridge Auth., 41 F. Supp. 638 (D.N.H. 1941).

[80] AIA A201-1987, art. 7.9; EJCDC 1910-8 (1990), Exhibit GC-A to General Conditions of the Agreement Between Owner and Contractor.

3. Streamline the selection of arbitrators and/or reduce the arbitration panel to a single arbitrator. Three-panel arbitrations are often more difficult to schedule and require more time for presentation. Preselection of arbitrators, or creation of a select panel, before commencement of work can also facilitate arbitrator selection. Preselection avoids problems with competency and conflict of interest. If a select panel is used, final selection may be random, at the arbitration administrator's discretion, or based upon each party's right to strike some of the panel members.
4. Reduce or eliminate discovery. If the arbitration occurs shortly after the discovery of the differing site condition, the parties should have relatively similar access to information. To reduce surprise, require exchange of any documents to be used during the arbitration and exchange of expert witness reports. Live testimony of experts should be eliminated, or experts should be precluded from expressing opinions not stated in their reports.
5. Consider placing time limits on presentations. This reduces the tendency to include redundant and peripheral issues.
6. Require arbitration to continue even if third parties cannot be joined. Under some state procedures, an arbitration can be stayed or enjoined if third parties cannot be joined,[81] and some jurisdictions do not permit consolidation of joint arbitrations when the contracts contain nonjoinder clauses.[82] Because the owner may desire to have its consultants joined with the owner-contractor arbitration, the owner-consultant agreements should contain arbitration provisions that mirror the owner-contractor agreement and that explicitly permit joinder.

Arbitration variations. *Hi/lo arbitration* is used to resolve differences between the final settlement positions of the parties. Once each party has made its final offer, the matter is arbitrated. The arbitrator is not allowed to award more than the claimant's lowest demand nor less than the respondent's highest offer. This guarantees claimant a favorable award while limiting the respondent's exposure.

Because the parties in hi/lo arbitration have relatively little at risk, it is possible to compress the arbitration proceedings. Often, the time allowed for presentation of evidence is sharply curtailed, as is the use of live testimony. The entire arbitration may be reduced to expert reports and written briefs.

Baseball arbitration is also based upon the last offer and demand. The award, however, is quite different. The arbitrator must award either the last demand or the last offer—nothing else. The arbitrator chooses between the two by determining whether the award is closer to the offer or the demand.

---

[81] *See, e.g.,* Cal. Civ. Proc. Code § 1281.4 (West 1982); Volt Info. Sciences v. Board of Trustees, 489 U.S. 468, 109 S. Ct. 1248, 1255 (1989).

[82] Del E. Webb Constr. v. Richardson Hosp. Auth., 823 F.2d 145 (5th Cir. 1987).

Baseball arbitration forces the parties to accurately assess the value of their positions. If a defendant offers too little, it runs the risk that the claimant's demand, though excessive, is closer to the arbitrator's determination. Similarly, if the claimant demands too much, the defendant's last offer may be awarded.

Nonbinding arbitration. Nonbinding arbitration is used when the parties desire a ruling on the issues and either do not want, or cannot agree to, be bound by the ruling. Public agencies, for instance, are often prohibited from engaging in binding arbitration. Other than enforceability, the mechanisms of nonbinding arbitration are very similar to arbitration. The United States Corps of Engineers has published a pamphlet describing its experience with nonbinding arbitration.[83]

Mediation. Mediation is the simplest of alternative dispute resolution devices. It often requires the least preparation, is the least time consuming, and can be arranged on short notice. Mediation is also effective. DPIC Companies, an insurer of design professionals, reports that 74 percent of cases accepted into mediation are resolved.[84]

Properly managed, mediation increases communication between the parties. Because it focuses on the interests of the parties, it is less likely to damage business relationships. Mediation can even be used to build business relationships because it allows the participants to show that they understand and are concerned with, their opponent's problems. Even if the mediation is unsuccessful, the parties will have a better understanding of their respective positions, which will streamline further proceedings.

There are two criticisms of mediation. First, mediation is nonbinding. Especially when negotiations have been hostile, parties believe that nonbinding dispute resolution wastes time and money because they assume their opponent will automatically reject a "reasonable" resolution. Second, parties believe that mediation, which often occurs before discovery, cannot result in a thoughtful resolution of a factually complex problem. The parties think they cannot settle because they lack sufficient information. Experience has shown that neither criticism is valid.

Mediation is particularly appropriate when negotiations have been hostile. It allows the parties to draw back from conflict and reestablish principled methods of dispute resolution and communication. In contrast, binding dispute resolution mechanisms encourage the parties to continue attacking each other's positions. This destroys, rather than enhances, business relationships.

---

[83] IWR Pamphlet-90-ADR-P-2.

[84] ASFE, Association of Firms Practicing in the Geosciences, Alternative Resolution for the Construction Industry 29 (1988).

Mediation is also a remarkably effective evaluation tool even when employed before discovery. The parties generally underestimate their ability to evaluate their disputes.

Minitrial. The minitrial is one of the best known alternative dispute resolution techniques[85] and has received considerable publicity and acclaim.[86] It is particularly suited to resolving larger, factually complex disputes. The United States Army Corps of Engineers actively promotes the use of minitrials and has issued a pamphlet describing its minitrial procedures and containing sample minitrial agreements.[87] The AAA also has sample minitrial procedures.[88]

Like most alternative dispute resolution mechanisms, minitrial procedures may be modified to suit the particular needs of the parties. The goal of the minitrial is to quickly educate senior decision makers concerning the strengths and weaknesses of the claims and defenses. In essence, a minitrial consists of an abbreviated presentation of each party's case before senior managers of the parties. A neutral advisor facilitates the conduct of the hearing and settlement discussions but does not make any awards or determinations. The neutral advisor may issue advisory opinions in the minitrial if the parties wish.

Disputes review board. A disputes review board, also known as a *standing arbitration panel,* is a standing panel that determines disputes between owner and contractor as they occur. Although mostly used on larger projects, disputes review boards have distinct advantages for differing site conditions claims. First, because the panel already exists, hearings can take place very quickly. Second, because the disputes review board is acquainted with the project, relatively little hearing time need be spent educating the panelists. Third, because the panel members are selected for their knowledge and expertise, the parties have considerable confidence in the panel's determinations. Finally, because the panel will potentially hear other disputes as well, it is in the parties' interest to avoid making statements or taking positions that could damage the party's credibility. Claims tend to be more reasonable.

§ 1.7 Insurance

Differing site conditions are not ordinarily an insurable risk. Hence, insurance is of limited utility in protecting against the risk of differing site conditions. In

---

[85] In a study of 114 companies, minitrials were the most commonly used alternative dispute resolution mechanisms (39%). Mediation was the second most common method (17%).

[86] *See Let's See More Minis,* Engineering News Record Dec. 3, 1987 and *Alternative Dispute Resolution—Opening Doors to Settlement,* Engineering News Record, Dec. 3, 1987, at 28.

[87] IWR Pamphlet-89-ADR-P-1.

[88] American Arbitration Association, Mini-Trial Procedures, AAA-148-5M-12/86.

some instances, however, insurance may be useful in resolving a differing site conditions claim after it has occurred. For example, if hazardous materials are discovered during excavation, the cost of remediation may be covered by liability insurance maintained by the owner or prior owners. Recent policies, however, have attempted to exclude this coverage. If the hazardous materials should have been discovered during the site investigation, the engineer's errors and omissions coverage may also pay a portion of the remediation costs.

Coverage may also exist for consequential damage related to the differing site condition. For instance, if an excavation collapse injures a third party's property, coverage may exist for this damage. It may also be possible to obtain reimbursement for stabilizing a situation threatening third-party property.

In general, however, one should not assume that insurance will play a direct role in avoiding differing site conditions claims.

CHAPTER 2

CONTRACT ALLOCATION OF RISK OF DIFFERING SITE CONDITIONS

John E. Beard III

John E. Beard III is a partner in the law firm of Kirkpatrick & Lockhart, where, from its Pittsburgh, Pennsylvania office, he specializes in counseling, general representation, and litigation relating to the real estate and construction industry, government contracts and environmental law. Mr. Beard received his J.D. from Ohio Northern University College of Law. His experience includes representation of construction contractors and engineering firms in a variety of litigation matters before state and federal courts and claims boards, and domestic and international arbitral tribunals. A member of the Public Contract Law Section and the Forum Committee on the Construction Industry of the American Bar Association, the Construction Litigation Committee of the Civil Litigation Section of the Pennsylvania Bar Association, for which he has been a contributing author and speaker, Mr. Beard is also a member of the Pennsylvania bar.

§ 2.1 History and Purpose

Typically, a modern construction contract will allocate, through a contract clause or clauses, the risk of differing site conditions. This chapter reviews different contract provisions that allocate that risk. In addition, when site conditions cannot be defined with reasonable accuracy, alternative methods to conventional contracting are examined.

The allocation of the risk of differing site conditions can vary substantially from contract to contract, depending upon the intent of the parties. Before the advent of the modern differing site conditions clause[1] (and today when such a clause is not used), the burden of differing site conditions rested primarily with the contractor. To recover for an unforeseen site condition in the absence of the differing site conditions clause, a government contractor, for example, was required to prove both a material variation between the condition encountered and the condition expected and that the government had misrepresented the condition by either knowingly or negligently misrepresenting a fact or by failing to disclose information it had a duty to disclose.[2]

As a practical matter, bidding subsurface work under such contract terms has a large speculative factor. The risk of the unknown must be accounted for by the contractor through bid contingencies or exhaustive site investigation. The absence of a contract mechanism for a price adjustment and the requirement of proof of government culpability favors litigation over responsible and contemporaneous problem solving and negotiation.

The modern differing site conditions clause was developed in the public works contracting arena, including projects such as sewers, bridges, roads, tunnels, and dams. The cost and difficulty of these construction projects often depend substantially on the characteristics of the subsurface, the nature and availability of naturally occurring materials, and the presence or absence of water.

The modern differing site conditions clause was intended to lessen the speculative character of site work by allowing the contractor to rely upon and bid site conditions that were indicated by the contract documents or that were normal for the work. By providing for notice and a contract remedy (an equitable adjustment) in the event of a material variation from the conditions expected, the contracting agency and the contractor were encouraged to resolve the issue within the contract. These clauses, when coupled with adequate site information, were intended to promote uniform and conservative bidding and thus lower bids and final costs to a contracting agency properly characterizing site conditions. For the contractor, the clauses ease the contractor's burden of risk assessment through either large bid contingencies or exhaustive site investigation. The clauses recognize, as a practical matter, that owners have access to geotechnical expertise and time in which to plan their projects, while contractors usually are not geotechnical experts and usually have limited time and ability to perform their own site investigations before bid.[3]

---

[1] The federal differing site conditions clause and the American Institute of Architects (AIA) concealed conditions clause are set forth in their entirety in § **2.2**.

[2] *See* Midland Land & Improvement Co. v. United States, 58 Ct. Cl. 671, 683–84 (1923), *aff'd,* 270 U.S. 251 (1926); C.W. Blakeslee & Sons v. United States, 89 Ct. Cl. 226, 250–51 (1939), *cert. denied,* 309 U.S. 659 (1940).

[3] *See* Servidone Constr. Corp. v. United States, 19 Cl. Ct. 346 (1990), *aff'd,* 931 F.2d 860 (Fed. Cir. 1991); J.F. Shea Co. v. United States, 4 Cl. Ct. 46 (1983); Foster Constr. C.A. v. United States, 435 F.2d 873 (Ct. Cl. 1970).

§ 2.2 Standard Clauses

Federal government contract clause. The standard federal government clause for differing site conditions applicable to fixed-price construction contracts provides:

> (a) The Contractor shall promptly, and before the conditions are disturbed, give a written notice to the Contracting Officer of
>
> (1) subsurface or latent physical conditions at the site which differ materially from those indicated in this contract, or
>
> (2) unknown physical conditions at the site, of an unusual nature, which differ materially from those ordinarily encountered and generally recognized as inhering in work of the character provided for in the contract.
>
> (b) The Contracting Officer shall investigate the site conditions promptly after receiving the notice. If the conditions do materially so differ and cause an increase or decrease in the Contractor's cost of, or the time required for, performing any part of the work under this contract, whether or not changed as a result of the conditions, an equitable adjustment shall be made under this clause and the contract modified in writing accordingly.
>
> (c) No request by the Contractor for an equitable adjustment to the contract under this clause shall be allowed, unless the Contractor has given the written notice required; provided, that the time prescribed in (a) above for giving written notice may be extended by the Contracting Officer.
>
> (d) No request by the Contractor for an equitable adjustment to the contract for differing site conditions shall be allowed if made after final payment under this contract.[4]

Standard American Institute of Architects (AIA) clause. Not dissimilar to the federal clause, AIA General Condition 12.2.1. Concealed Conditions provides:

> Should concealed conditions encountered in the performance of the Work below the surface of the ground or should concealed or unknown conditions in an existing structure be at variance with the conditions indicated by the Contract Documents, or should unknown physical conditions below the surface of the ground or should concealed or unknown conditions in an existing structure of an unusual nature, differing materially from those ordinarily encountered and generally recognized as inherent in work of the character provided for in this Contract, be encountered, the Contract Sum shall be equitably adjusted by Change Order upon claim by either party made within twenty days after the first observance of the conditions.

---

[4] 48 C.F.R. § 52.236-2 (1991).

§ 2.3 —Elements of Standard Clauses

As explained in more detail in subsequent chapters, the standard clauses provide for an equitable adjustment for two distinct types of differing site conditions.

Type I consists of unforeseeable subsurface or latent conditions at the site that differ materially from those indicated in the contract. Entitlement to a Type I claim requires proof of (1) the express or implied conditions indicated in the contract and (2) an encountered condition that differs materially from the conditions indicated. The basis of comparison for a Type I claim is the encountered condition compared with contract indications of site conditions.[5] The threshold issue of what is "indicated" in the contract is a matter of law, with the tribunal placing itself in the shoes of a "reasonable and prudent" contractor for purposes of contract interpretation.[6]

Type II consists of unknown and unusual conditions at the site that differ materially from those ordinarily encountered and generally recognized as inhering the work. Type II cases arise when the contract does not contain indications of the site conditions encountered.[7] The frame of reference for comparison of encountered conditions is the "normal" condition for the area and type of work to be performed.[8] In addition to proof of unusual conditions, the contractor must also prove that the encountered condition was unknown, that is, that a reasonably prudent contractor would not have anticipated the condition from the contract indications or a reasonable site investigation or inquiry.[9]

Both Type I and Type II claims require that the conditions encountered "differ materially" from the reference conditions, whether "indicated" by contract or "normal." The contractor has the burden of proof on these issues, which are questions of fact and often involve expert testimony.[10] Other evidence that has been instrumental in showing a material difference in the condition

---

[5] P.J. Maffei Bldg. Wrecking v. United States, 732 F.2d 913, 916 (Fed. Cir. 1984); Foster Constr. C.A. v. United States, 435 F.2d 873, 881 (Ct. Cl. 1970).

[6] *See* P.J. Maffei Bldg. Wrecking v. United States, 732 F.2d 913, 917 (Fed. Cir. 1984).

[7] Contract silence can also give rise to a Type I claim when the failure to specify an element of work may give rise to an inference that site conditions requiring that work would not be encountered. *See, e.g.,* Foster Constr. C.A. v. United States, 435 F.2d 873, 889–92 (Ct. Cl. 1970) (failure to specify a "seal" or seal class concrete for cofferdam construction gave rise to inference that subsurface conditions would permit excavation in the dry, without a seal); CCM Corp. v. United States, 20 Cl. Ct. 649 (1990) (failure to specify a decontamination procedure was indication in waterproofing contract of absence of tar in previously installed waterproofing).

[8] *See* CCM Corp. v. United States, 20 Cl. Ct. 649 (1990); Servidone Constr. Corp. v. United States, 19 Cl. Ct. 346, 366 (1990), *aff'd,* 931 F.2d 860 (Fed. Cir. 1991).

[9] Servidone Constr. Corp. v. United States, 19 Cl. Ct. 346, 371 (1990) (contract indications and site investigation); Stock & Grove, Inc. v. United States, 493 F.2d 629, 635 (Ct. Cl. 1974) (site investigation).

[10] Arundel Corp. v. United States, 515 F.2d 1116, 1123 (Ct. Cl. 1975); Charles T. Parker Constr. Co. v. United States, 433 F.2d 771, 778 (Ct. Cl. 1970).

encountered includes proving that (1) a change was required in the method of performance from that contemplated by the contract;[11] (2) the indicated method of performance became cost prohibitive;[12] or (3) a design change was required by encountered conditions.[13]

The contracting agency is not required through the differing site conditions clause to describe site conditions with precision.[14] This is especially so when a reasonable site investigation would reveal the true condition.[15]

Notice. The standard clauses require that notice be given contemporaneously with the discovery of the differing site condition. The federal clause further requires that the contracting officer promptly investigate the encountered conditions. The notice requirement is primarily a litigation avoidance mechanism that has practical purposes and is not intended to be punitive. The purposes served are to allow the government the opportunity to evaluate undisturbed conditions and to exercise control over the resolution of the problem.[16]

Whether requisite notice has been given is a question of fact. The requirement of a written notice has been waived when the government in fact has actual or constructive notice of the conditions and difficulties encountered by the contractor and when no prejudice resulted from a lack of timely written notice.[17]

Scope of differing site conditions clauses. In addition to subsurface claims, the standard differing site conditions clause is sufficiently broad to include other types of latent conditions. The clause is especially pertinent to repair or remodeling contracts.[18]

---

[11] *See, e.g.*, Vann v. United States, 420 F.2d 968 (Ct. Cl. 1970) (spoil bottom rather than rock bottom prevented either drilling or driving of piles); J.F. Shea Co., IBCA No. 1191-4-78, 82-1 B.C.A. (CCH) ¶ 15,705 (1982) (rock fallout required redesign of approved tunnel machine).

[12] *See, e.g.*, Kaiser Indus. Corp. v. United States, 340 F.2d 322 (Ct. Cl. 1965) (excessive waste from indicated quarry made cost of performance unreasonable and prohibitive).

[13] *See, e.g.*, Foster Constr. C.A. v. United States, 435 F.2d 873, 892 (Ct. Cl. 1970) (bridge redesign required by subsurface conditions).

[14] Pacific Alaska Contractors, Inc. v. United States, 436 F.2d 461, 469 (Ct. Cl. 1971) (location and productiveness of borrow pits); Charles T. Parker Constr. Co. v. United States, 433 F.2d 771, 778–79 (Ct. Cl. 1970) (relative hardness of rock).

[15] Clark v. United States, 5 Cl. Ct. 447, 449 (1984), *aff'd*, 770 F.2d 180 (Fed. Cir. 1985) (claimed unforeseen conditions would have been "readily apparent" from reasonable site inspection).

[16] Brechan Enters., Inc. v. United States, 12 Cl. Ct. 545, 549 (1987).

[17] *See* Central Mechanical Constr., ASBCA No. 29,431, 85-2 B.C.A. (CCH) ¶ 18,061 (1985).

[18] *See, e.g.*, Yamas Constr. Co., ASBCA No. 27,366, 87-2 B.C.A. (CCH) ¶ 19,695 (1987) (Type II claim asserting that uneven concrete floors resulted in excessive glue removal in carpet replacement contract); Latham Co. v. United States, 20 Cl. Ct. 122 (1990) (claim that "built-up" roof indicated was different than "double-poured" roof encountered during roof repair contract; Midwest Indep. Painting, ASBCA No. 35194, 90-3 B.C.A. (CCH) ¶ 23,094 (1990) (amount of rotted wood greatly exceeded amount indicated in contract, thereby increasing scope of painting contract); Roberts Constr. Co., ASBCA No. 33949, 87-2 B.C.A. (CCH)

Generally, weather is not a risk that can be shifted to the government through the differing site conditions clause.[19] Likewise, a change in government regulations or an act in the government's public and general capacity as sovereign does not furnish a basis for a changed conditions claim or government liability as a contractor.[20]

§ 2.4 Types of Provisions Relating to Differing Site Conditions

Clauses that attempt to foreclose recovery. The standard differing site conditions clauses are by no means used universally. In particular, contracts with many state agencies and municipalities are often drafted to prohibit differing site condition claims. These provisions are enforced on the basis that they are necessary to protect public entities and public funds.[21] From a bidder's perspective, these clauses intend to place the risk of differing conditions on the contractor, and that risk must be evaluated through bid contingencies and on-site investigation.

For example, until recently, the Pennsylvania Department of Transportation did not include the differing site conditions clause in its specifications and had a standard contract provision that stated:

> The Contractor further covenants and warrants that he has had sufficient time to examine the site of the project; that he has examined the site of the project; that he has had sufficient time to examine the site of the project to determine the character of the subsurface materials and conditions to be encountered; that he is fully aware and knows of the character of the subsurface materials and conditions to be encountered; and that he has based the within contract prices on

---

¶ 19,897 (1987) (Type II claim when tunnel location for replacement refrigeration lines limited access due to obstructions and insufficient size); P-B Eng'g Co., ASBCA Nos. 39041, 39042, 39043, 90-2 B.C.A. (CCH) ¶ 22,686 (1990) (Type II claim asserting that discovery of reinforcement in concrete basement floors was an unusual condition for waterproofing contract).

[19] *See, e.g.*, Hardeman-Monier-Hutcherson v. United States, 458 F.2d 1364 (Ct. Cl. 1972); Tombigbee Constructors v. United States, 420 F.2d 1037 (Ct. Cl. 1970); *cf.* Dillon v. United States, 156 F. Supp. 719 (Ct. Cl. 1957) (entitlement based on impact of severe drought on hay supply contractor).

[20] *See, e.g.*, Amino Bros. Co. v. United States, 372 F.2d 485 (Ct. Cl. 1967), *cert. denied*, 389 U.S. 846 (1967) (government's release of water from dam resulted in washout of contractor's crossing); Warner Elec., Inc., VABCA No. 2106, 85-2 B.C.A. (CCH) ¶ 18,131 (1985) (change in PCB regulations).

[21] *See, e.g.*, Wiechmann Eng'rs v. State, 107 Cal. Rptr. 529 (Ct. App. 1973); Miami Dade Water & Sewer Auth. v. Inman, 402 So. 2d 1277 (Fla. Dist. Ct. App. 1981), *petition denied*, 412 So. 2d 466 (Fla. 1982); W.H. Lyman Constr. Co. v. Village of Gurnee, 403 N.E.2d 1325 (Ill. App. Ct. 1980).

> his own independent examination and investigation of the project site, subsurface materials and conditions and has not relied on any subsurface information furnished to him by the Commonwealth of Pennsylvania, Department of Transportation, its agents or its consultants.[22]

To justify entitlement to recovery in the face of such provisions, a contractor is required to demonstrate fraud or misrepresentation.[23] The elements required for entitlement were set forth in *Acchione & Canuso, Inc. v. Commonwealth Department of Transportation*[24] and may be summarized as follows:

1. The contracting agency makes a positive representation or specification of conditions of the work
2. The representation is material
3. The contractor has no reasonable means of making an independent investigation of the conditions or representations
4. The representations prove to be false or misleading either due to actual misrepresentation or gross mistake or arbitrary action by the contracting agency or its agents
5. As a result of the misrepresentation, the contractor suffers financial harm.

Clauses such as Pennsylvania's, where still found, are arguably contrary to federal public policy and should be objected to by bidders when the project is financed in whole or in part by federal funds. Pursuant to the 1987 amendments to the Federal Surface Transportation Act,[25] Congress required that construction contracts for federally aided highways contain certain standard clauses, including a clause that equitably addressed differing site conditions.[26] By regulation, the Federal Highway Administration now requires the inclusion of the standard federal differing site conditions clause in federally aided highway contracts.[27] Likewise, the federal Environmental Protection Agency (EPA) requires the clause in contracts for federally aided pollution abatement and control projects.[28]

Exculpatory clauses. Subsurface information, when presented in or with a contract, is often accompanied by exculpatory language indicating that such information should not be relied upon or may not be representative of conditions

---

[22] Pennsylvania Dep't of Transportation, Standard Forms.

[23] *See* Pennsylvania Turnpike Comm'n v. Smith, 350 Pa. 355, 39 A.2d 139 (1944); *overruled by* Lichtenstein v. Pennsylvania Turnpike Comm'n, 398 Pa. 415, 158 A.2d 461 (1960).

[24] 501 Pa. 337, 461 A.2d 765, 768 (1983).

[25] Pub. L. No. 100-17, 101 Stat. 150 (1987).

[26] 23 U.S.C. § 112(e) (Supp. 1991).

[27] 23 C.F.R. § 635.131(a)(1) (1991).

[28] 40 C.F.R. § 35, subpt. E, app. C-2 (1991).

to be encountered. For example, the Pennsylvania Department of Transportation, although now required to include the differing site conditions clause in its specifications, also includes the following language:

> Subsurface soil and geological information indicated is based upon soundings, dug test pits, and/or test borings. Such information concerning the character of subsurface material is of a preliminary nature and has been obtained for the exclusive use of the Department to aid in the project design. Do not consider any of the data supplied or made available to you by the Department or its authorized agents as positive representations of any of the conditions that you will encounter in the field. This information is not a part of the plans, proposal, or contract and is not to be considered a basis for computation of the unit prices used for bidding purposes. There is no expressed or implied agreement that the depth or character of the material has been correctly indicated or that uniformity of material exists between the explored locations. *The bidder is not to rely on this information, but is to assume the possibility that conditions affecting the cost and/or quantities of work to be performed may differ from those indicated.*[29]

There exists an apparent conflict between a provision that directs bidders to assume in their bid the possibility of differing site conditions and the standard differing site conditions clause. Indeed, the public policy and benefits to the contracting agency underlying the differing site conditions clause would appear to be negated by the inclusion in the same contract of the cited provision. The obvious import of the provision is to relieve the contracting agency of liability for differing site conditions.

On the federal level, these apparent conflicts have been resolved in favor of the differing site conditions clause. The rationale of these holdings is that exculpatory provisions in the invitation for bids or the specifications, irrespective of their breadth, cannot be construed to negate a mandatory contract provision such as the differing site conditions clause. Thus, a variety of general disclaimers—that the logs are not guaranteed and are not representations and that bidders must draw their own conclusions—have been rejected as in conflict with the differing site conditions clause.[30]

Core borings have been recognized in the courts as the most specific and reliable indication of subsurface conditions, even when the borings were taken at some distance from the encountered condition or during a time unlikely to indicate the presence of subsurface water.[31] Specific warnings or disclaimers contained in the drawings and specifications relating to the reliability of core

---

[29] Commonwealth of Pa., Dep't of Transp. 408/90 § 102.05 (emphasis added).

[30] *See* Foster Constr. C.A. v. United States, 435 F.2d 873, 888 (Ct. Cl. 1970); United Contractors v. United States, 368 F.2d 585, 598 (Ct. Cl. 1966); Fehlhaber Corp. v. United States, 151 F. Supp. 817, 825 (Ct. Cl.), *cert. denied,* 355 U.S. 877 (1957).

[31] Foster Constr. C.A. v. United States, 435 F.2d 873, 888 (Ct. Cl. 1970); Woodcrest Constr. Co. v. United States, 408 F.2d 406, 409–11 (Ct. Cl. 1969), *cert. denied,* 398 U.S. 958 (1970); United Contractors v. United States, 368 F.2d 585, 598 (Ct. Cl. 1966).

borings, the suitability, availability, or nature of material, or the presence of water are not necessarily dispositive of differing site condition claims, but must be read in harmony with the contractor's view of all indications, the contractor's opportunity through site inspection to reveal contrary indications, the nature of the inspection made, and the conditions actually encountered. The indications provided by core borings are not negated by undefined or unexplained general cautionary language.[32] However, specific interpretive language will be given full effect in determining contract indications.[33]

Estimated quantity provisions. In public work, it is not unusual to submit unit prices for particular bid items based upon estimated quantities of work. Often estimated quantities are accompanied by contract clauses known as the *approximate quantity* and the *variations in quantity* clauses. These clauses provide that the quantities identified are approximates only and are not to be relied upon and that variations from the estimated quantity in the actual work performed will not be the basis for an adjustment in contract unit price. More recently, the clauses have provided that only a "significant" change justifies a change in unit price. Typical of such a clause is the following:

> The term "significant change" applies only to the following circumstances:
>
> - When the character of the work as altered differs materially in kind or nature from that involved or included in the original proposed construction, or
> - When an item or work is increased in excess of 125% or decreased below 75% of the original contract quantity. Any allowance for an increase in quantity applies only to that portion in excess of 125% of original contract item quantity or, in case of a decrease, below 75% to the actual amount of work performed.[34]

Both estimated quantities and the variations in quantity clause have a bearing on the differing site conditions clause. The Court of Claims has rejected the proposition that a significant variation from an estimated quantity, without more, constitutes a differing site condition.[35] Estimated quantities can serve as the sole basis for a changed condition claim when the change in quantities could not have been foreseen, such as when the contractor before bid had no

---

[32] *See, e.g.,* Kaiser Indus. Corp. v. United States, 340 F.2d 322 (Ct. Cl. 1965) (warning of presence of "excess fines" in quarry could not be construed to mean that quarry was essentially unusable); United Contractors v. United States, 368 F.2d 585, 598 (Ct. Cl. 1966) (warning of "high ground water" did not negate boring information indicating little water was present at elevation of sewer installation).

[33] *See, e.g.,* Jefferson Constr. Co. v. United States, 364 F.2d 420 (Ct. Cl. 1966), *cert. denied,* 386 U.S. 914 (1967) (warning that borrow pits were probably insufficient to provide all fill for a levee was effective to bar contractor's claim for additional cost for fill).

[34] Commonwealth of Pa., Dep't of Transp., 408/90 § 110.02(d).

[35] Perini Corp. v. United States, 381 F.2d 403, 410 (Ct. Cl. 1967).

opportunity to verify the quantities.[36] Moreover, estimated quantities may provide contractors with indications of subsurface conditions and may be relied upon as such, even when the estimated quantities are accompanied by a disclaimer that they were provided solely for bid comparison purposes.[37]

For example, in *J.F. Shea Co. v. United States,*[38] the contractor bid and obtained a contract to build an aqueduct. As support for a tunnel portion of the work, the contracting agency estimated 2,190,000 pounds of steel rib supports and 140,000 linear feet of rock bolt supports. The contractor asserted that the estimated quantities, together with the core borings and other information, indicated that subsurface conditions would be consistent with 50 percent of the tunnel being supported by rock bolts. In fact, encountered subsurface conditions required that the entire tunnel be supported by steel ribs.

The court held that the estimated quantities, even when accompanied by a disclaimer, were contract "indications" to be considered in connection with all other information provided or available on subsurface conditions. The estimated quantities served as a confirmation of the contractor's understanding of the subsurface conditions and supported a Type I claim under the differing site conditions clause for its additional cost caused by the differing subsurface condition.

Another instructive case is *Perini Corp. v. United States,*[39] in which the contractor bid a high unit price on a bid item for water pumping, based upon the contractor's assessment that the contract estimated quantity for the work item was substantially less than would be required by encountered conditions. Actual quantities of water greatly exceeded the estimated quantities, due in part to the government's failure to order grouting of underground cavities. In a rather unique action, the government sought a reformation of the contract unit price, urging that the contractor was receiving a bonanza by reason of differing site conditions. The contractor demonstrated that through the information available at the site, it knew and the government should have known that the estimated quantity was erroneously low and that large amounts of water would be encountered. The court held that the differing site conditions clause did not relieve the government of its error any more than it would relieve a contractor of a bad estimate.

---

[36] *See, e.g.,* Schutt Constr. Co. v. United States, 353 F.2d 1018 (Ct. Cl. 1965) (overrun of 39% on unverifiable estimate of land acreage to be cleared); Chernus v. United States, 75 F. Supp. 1018 (Ct. Cl. 1948) (overestimate of 37% in cubic yards of excavation caused contractor to obtain the wrong equipment).

[37] J.F. Shea Co. v. United States, 435 F.2d 873 (Ct. Cl. 1970). *See also* Continental Drilling Co., ENGBCA No. 3,455, 75-2 B.C.A. (CCH) ¶ 11,541 (1975) (small estimated quantity for casing for core drilling contract served as contract indication that subsurface would be well cemented and tight in place).

[38] 435 F.2d 873 (Ct. Cl. 1970).

[39] 381 F.2d 403, 410 (Ct. Cl. 1967).

The Court of Claims, in construing the variations in quantities clause with the differing site conditions clause, has consistently held that unit prices bid, even for unclassified excavation, do not control the reimbursement of a contractor for work that is changed in character and reimbursable because of a differing site condition. The rationale of these holdings is that the unit price bid was for work of a different character than that actually encountered.[40]

Classified excavation. When it reasonably can be expected that site conditions of different characteristics will be encountered on a project, the better contracting practice is for the owner to classify into types the expected subsurface conditions. This is commonly known as *classified excavation* as contrasted with unclassified excavation and, for example, is commonly found in tunneling contracts. Unit prices can then be solicited on the different types of conditions to be encountered when the project will be closely monitored for quantities. Alternatively, having indicated the various types of conditions to be encountered, the contracting agency can solicit a composite, or lump sum, bid based upon an estimated quantity of all types of work that are present.[41]

In conjunction with identifying the types of material to be encountered, good contracting practice would also justify indication in the contract of the expected behavior of the identified types of conditions. A successful Type I claim can be premised upon unexpected behavior of indicated subsurface conditions.[42] When describing behavioral conditions, the owner should do so expressly in terms a contractor is likely to understand.

For example, in *Shank-Artukovich v. United States,*[43] although the types of excavation were identified, the tunneling contractor's claim was premised on two unexpected behaviors known as *caving ground* and *running ground.* The government defended on the basis that such behaviors were indicated by the core borings and technical specifications. The court concluded:

> It is clear that the government did not expressly include the terms "caving ground" or "running ground" in the contract. They stopped short of using express terms to define the behavior that they actually seemed to expect, if

---

[40] *See, e.g.,* Morrison-Knudsen Co. v. United States, 397 F.2d 826, 843 (Ct. Cl. 1968); United Contractors v. United States, 368 F.2d 585, 598 (Ct. Cl. 1966).

[41] In such a case, a Type I claim can be predicated upon a material change from the estimated quantities of the mixed types of work. *See* Peter Kiewit Sons' Co. v. United States, 74 F. Supp. 165 (Ct. Cl. 1947).

[42] *See, e.g.,* Shank-Artukovich v. United States, 13 Cl. Ct. 346 (1987), *aff'd,* 848 F.2d 1245 (Fed. Cir. 1988); J.F. Shea Co., IBCA No. 1191-4-78, 82-1 B.C.A. (CCH) ¶ 15,705 (1982); Continental Drilling, ENGBCA No. 3,455, 75-2 B.C.A. (CCH) ¶ 11,541 (1975).

[43] 13 Cl. Ct. 346 (1987), *aff'd,* 848 F.2d 1245 (Fed. Cir. 1988).

> Dr. Goodman's expertise is to be imputed to the government. What the government seems to have done was to include in the specifications language which could be construed by an expert to mean that caving or running ground should have been expected. In this manner the government avoids high bids and retains a sound argument for inferring caving and running behavior from the specifications. This occurs to the court to be a misuse of the differing site conditions clause.
>
> If the government expected caving or running ground, it should have expressly said so. It expressly indicated swelling, squeezing, and raveling ground. If the inference to caving or running ground was as clear as the government's witnesses maintain, then the terms should have been expressly included in the contract and not used as an opportunity to sandbag the contractor.[44]

As indicated in the foregoing discussion, the best approach to allocating the risk of differing subsurface conditions is to place the risk on the party that is in the best position to manage the risk. As respects an owner who has access to geotechnical expertise and time to plan a project, the better practice is to describe subsurface conditions not only with geotechnical data but also with the conclusions drawn from that data in terms a contractor is likely to understand. By giving adequate information, the owner should be rewarded with competitive bids from reliable contractors.

As to the contractor, the importance of an adequate site investigation cannot be overstated. The case is rare in which a contractor is permitted recovery absent an adequate pre-bid investigation.[45] When subsurface conditions are a material part of the contract, access by the contractor to geotechnical expertise in advance of the bid can sometimes be critical to evaluation of contract indications and submission of a competitive bid.

When unexpected conditions are encountered, the standard differing site conditions clause allows the parties to manage the condition in real time, from an engineering and construction perspective rather than from a legal perspective. Blatantly unfair contract provisions that place the risk on the contractor, irrespective of its ability to manage and foresee that risk, simply breed litigation, the transactional cost of which is best avoided.

---

[44] 13 Cl. Ct. at 355.

[45] Such a rare case occurred in Servidone Constr. Corp. v. United States, 19 Cl. Ct. 346 (1990), *aff'd*, 931 F.2d 860 (Fed. Cir. 1991). The court found that the contractor had conducted an imprudent pre-bid investigation, especially in view of its inexperience in the geographic area of the project. As a result, the contractor's bid was erroneous. The court nevertheless found that a changed condition existed from that which a prudent pre-bid investigation would have revealed and allowed recovery for the difference between reasonable cost and cost actually incurred. On the issue of damages, the court considered the other bids received on the project.

§ 2.5 Approaches for Undefined Site Conditions: Environmental Remediation Contracts

It is hoped that advances in geotechnical science, contractor experience, and legal jurisprudence have combined to take some of the gamble out of subsurface operations on conventional projects. However, a multitude of new claims relating to differing site conditions can be expected to arise out of environmental remedial action contracts (RACs).

It is estimated that as many as 30,000 sites in the United States are contaminated to some extent by toxic chemicals.[46] Pursuant to the policy and remedies established under the Comprehensive Environmental Response, Compensation and Liability Act[47] (CERCLA or Superfund), the EPA, other government agencies such as the Departments of Defense and Energy, and private parties have been investigating sites and developing remedial designs for those sites. EPA purports to have completed cleanups of 63 sites and has a goal of completing cleanup of 200 sites by 1993 and 650 sites by 2000.[48] The work of the Departments of Defense and Energy in cleaning up government-owned property may exceed the EPA program. Over 60 percent of Superfund sites will be cleaned up by private industry.[49]

Conventional contracting presumes the existence of a relatively specific set of plans and specifications which in turn will allow costs to be estimated with reasonable certainty. Remediation of contaminated property involves substantial unknowns, despite exhaustive efforts in remedial investigation, waste characterization, and feasibility studies. Typically, despite exhaustive study, the character, location, and quantity of waste will be uncertain; known and unknown, surface or subsurface conditions have the potential to disperse the waste, with potential liability to contractor and owner. Thus, the methods of treatment or disposal may be diverse, unproven, complicated, or proprietary.

The complexity and nebulous character of the RAC work is coupled with an unclear legal or financial framework in which to perform the work. Issues

---

[46] *See Executive Summary of EPA Alternative Superfund Contracting Strategy Report,* 22 Env't Rep. (BNA) 1,505 (Oct. 4, 1991).

[47] Pub. L. No. 96-510 (1980) (codified at 42 U.S.C. §§ 9601–9675 (Supp. 1991)) [hereinafter CERCLA].

[48] 22 Env't Rep. (BNA) 2,140 (Jan. 17, 1992).

[49] *See Executive Summary of EPA Alternative Superfund Contracting Strategy Report,* 22 Env't Rep. (BNA) 1,505 (Oct. 4, 1991).

relating to liability,[50] indemnity,[51] insurance,[52] and cleanup standards[53] are unresolved and the subject of much public debate. Obviously, contractors should insist on the differing site conditions clause being included in RACs. Unforeseen environmental hazards are recognized as differing site conditions.[54]

Conventional contracting is certain to give rise to litigation such as that recently filed in *IT Corp. v. MOTCO Site Trust Fund.*[55] At issue is the cleanup of the MOTCO site, an 11-acre site about five miles north of Galveston, Texas. According to the complaint, the site was used for reclaiming waste containing polychlorinated biphenyls, styrene tars, copper mercuric chloride, mercury, and lead.

---

[50] CERCLA, at 42 U.S.C. § 9619(1) (1991), provides:

> Sec. 119. (a) Liability Of Response Action Contractors.—
>
> (1) Response Action Contractors.—A person who is a response action contractor with respect to any release or threatened release of a hazardous substance or pollutant or contaminant from a vessel or facility shall not be liable under this title or under any other Federal law to any person for injuries, costs, damages, expenses, or other liability (including but not limited to claims for indemnification or contribution and claims by third parties for death, personal injury, illness or loss of or damage to property or economic loss) which results from such release or threatened release.

This provision would appear to protect the response action contractor against liability under federal law for preexisting or threatened releases. The provision does not address state law claims. Moreover, when a contractor through subsurface operation or negligence causes a second release or disperses the preexisting release, the provision is of questionable applicability. *See* Moskowitz, *Superfund Contractor Indemnification,* 8 Envtl. L. Rep. (Envtl. L. Inst.) 1,033 (1990); *see also* Tanglewood E. Homeowners v. Charles-Thomas, Inc., 849 F.2d 1568 (5th Cir. 1988) (recognizing that a contractor's action could constitute a second release).

[51] Section 9619(e) of CERCLA, added in 1986, provides for government indemnification for response action contractors for negligence in the performance of carrying out response action activities, unless the conduct was grossly negligent or constituted intentional misconduct. The indemnification is discretionary and, to be eligible, the contractor must demonstrate "diligent efforts to obtain insurance coverage" and that the insurance is not available at a "fair and reasonable price." CERCLA at 42 U.S.C. § 9619(c)(4) (Supp. 1991). The scope of indemnification is a contract by contract negotiation process.

[52] The market for contractors' environmental protective liability insurance is quite restrictive. Policies are written on a claims-made basis. Thus, a remedial action contractor is likely to have substantial uninsured risk. EPA now maintains that insurance is available at fair and reasonable prices. *See* 22 Env't Rep. (BNA) 1,937 (Dec. 6, 1991).

[53] Often the objective of a RAC is unclear. EPA is required to clean up sites to "applicable" or "relevant and appropriate requirements" (ARARs). CERCLA, 42 U.S.C. § 9621 (Supp. 1991). The blueprint for achieving ARARs is a steady stream of regulations promulgated by EPA known as the National Contingency Plan. 40 C.F.R. § 300 (1991). However, in reality, EPA has substantial discretionary authority, and decision making proceeds on a site-by-site basis, with engineering and construction considerations being only one factor EPA must consider.

[54] *See, e.g.,* Charles G. Williams Constr., Inc., ASBCA No. 42246, 91-3 B.C.A. (CCH) ¶ 24,309 (1991) (asbestos).

[55] No. 91-3532 (S.D. Tex. Dec. 3, 1991).

For a contract price of $30 million, IT agreed to design, engineer, and build an incinerator and to destroy the waste through incineration. IT now claims that the subsurface for the incinerator foundations was seriously contaminated, requiring a revision to the proposed installation and erection scheme. IT also asserts that subsequent studies of the site reveal that the actual wastes located at the site were "nearly 100 percent different" from the conditions indicated in the trust's studies, thereby necessitating changes in the remediation plan and schedule. IT has submitted a revised proposal for $100 million dollars and ceased on-site work. The trust maintains that its studies were correct and that all data were given to IT.[56]

This embryonic litigation demonstrates the weakness in utilizing the conventional fixed-price approach for RACs. First, it may not be possible to define the work sufficiently to establish a firm fixed price. Secondly, it places the owner, its engineers, and geotechnical team in an adversarial position to the contractor, with each unexpected encounter resulting in a price negotiation. Other approaches should be considered.

Turnkey contracting. Under a turnkey or design-build approach, one firm or a consortium of firms would have site responsibility from investigation and design through cleanup. From an owner's perspective, this provides single-source responsibility and thereby eliminates the gap in both communication and liability between designer and contractor. The approach also has the advantage of bringing to bear in the design phase the contractor's perspective and expertise on such subjects as constructability, value engineering, cost, and selection of remedial techniques.

The perceived disadvantages of this approach are the absence of competition and higher cost. It is rare to find one firm with the expertise or capability to perform investigation, design, cleanup, and disposal. However, teaming arrangements can be encouraged and employed to provide initial competition.[57] Competitive proposals can be evaluated on hourly rates and fees, among other criteria. The selected prime contractor can subcontract on a competitive basis for expertise or services in areas where the prime contractor is perceived to be deficient. Competition can also be fostered by requiring the prime contractor to bid competitively to third parties portions of the work, when and as the work becomes clearly defined. Finally, budgeting and cost management by the owner can create cost consciousness for that portion of the work performed on other than a fixed-price basis.

Cost-plus approach. When portions of the work are ill defined or present a substantial risk of unanticipated conditions or results, a sound approach is for

---

[56] 22 Env't Rep. (BNA) 2,099 (Jan. 3, 1992).

[57] EPA has formed a task force to develop a program by September 1992 to team together contractors for work on Superfund sites. 22 Env't Rep. (BNA) 305 (June 7, 1991).

the owner and contractor to recognize that fact initially and to share the risk for that work through a cost reimbursement arrangement. Such arrangements may include features allowing the parties to fix a lump sum or unit price for the work at such time as the work becomes defined. Cost-plus features can be employed in conjunction with fixed prices for defined portions of the work. The key element is the recognition at the time of contracting that the work is ill defined.

Construction management. Construction management offers many of the advantages of the turnkey approach in that the contractor's expertise, capability, and experience are brought to bear early in the project, along with that of the engineers and geotechnical experts. The approach also lends itself to phased procurement and execution, and competitive bidding of work as and when it becomes defined.

Construction management may be particularly appropriate to private party cleanups. Such cleanups are often funded by numerous potentially responsible parties represented by various committees or subgroups. In such cases, either because none of the parties wishes to devote substantial effort to manage the project or because the parties have conflicting interests, it may be entirely appropriate to delegate substantial project management responsibilities to an independent third party.

Current EPA regulations requiring a remedial investigation and feasibility study, followed by agency action, a remedial design, followed by agency action, and finally site remediation lend themselves to the conventional contracting approach.[58] However, as procurement for actual remediation begins in earnest, the government is considering alternatives to conventional contracting in view of the broad and ill-defined scopes of work and liabilities inherent in many remediation projects.[59]

Private parties would be well advised to add a contractor's expertise at the earliest possible time to supplement their investigative and design teams. As a

---

[58] Under the approach mandated by CERCLA, specified activities must be accomplished to move a site from identification to cleanup. Those activities start with a remedial investigation/feasibility study (RI/FS), intended to identify a site work plan and several cleanup options. EPA then selects among the alternatives and prepares and publishes a "Record of Decision" (ROD) in which it announces the tentative choice among the RI/FS options. The ROD is followed by the detailed design of the general remedial concept, which includes the detailed engineering plans and site specifications. Once design is complete, a construction contract or contracts may be awarded. Recently, sites have been divided into "operable units" which allow the work to proceed on a phased basis. *See* CERCLA, at 42 U.S.C. § 9621 (Supp. 1991); National Contingency Plan, 40 C.F.R. §§ 300.430–.435 (1991).

[59] *See Executive Summary of EPA Alternative Superfund Contracting Strategy Report,* 22 Env't Rep. (BNA) 1,505 (Oct. 4, 1991).

practical matter, many private party Superfund actions are and will be settled based upon cost allocation agreements and risk assessments made, much like a contractor's bid, on site indications that will eventually prove to be inaccurate. The risk of differing subsurface conditions is more efficiently approached through initial engineering, planning, phased procurement, and reasonable risk allocation than through litigation.

CHAPTER 3

EFFECT OF NOTICE REQUIREMENTS ON DIFFERING SITE CONDITIONS CLAUSES

J. Mark Brewer*

J. Mark Brewer is a 1981 graduate of the University of Missouri School of Law—Kansas City and also holds a master of arts degree in international relations from the Cambridge University (England) School of International Relations, University of Southern California. Mr. Brewer spent four years on active duty as an assistant staff judge advocate and chief of military justice in the United States Air Force Judge Advocate General's Department, during which he was stationed in the United Kingdom and Florida. Since his days as a military prosecutor, Mr. Brewer's practice has been devoted largely to the litigation of construction-related disputes on behalf of general contractors and their sureties. He is named principal in the Houston, Texas, law firm of Brewer & Pritchard, P.C.

§ 3.1 Standard Notice Requirements

Differing site conditions clauses almost always require that notice be given to the other party on the discovery of a differing site condition. Construction documents usually require that notice to be in writing. This chapter focuses on the

*The author wishes to express his appreciation for the assistance of Patrick E. Gaas, an associate with Brewer & Pritchard, P.C., in preparing this chapter.

effect, if any, of the notice requirements contained either within the differing site conditions clause itself or elsewhere in the contract on the ability of the contractor to recover additional compensation for a differing site condition.

Even though most contract documents require *written* notice of the discovery of a differing site condition (by the contractor), courts have, more often than not, held in favor of the contractor whenever the contractor gives some form of notice, albeit deficient. These holdings are often based on the doctrines of waiver or estoppel. Occasionally, courts do rule against contractors on the same doctrines, holding that the contractor's failure to comply with the notice requirement is without excuse or justification. In essence, the results of any particular case will vary with the particular fact pattern, notice provision, and jurisdiction of the case. See §§ **3.3** and **3.4**.

This chapter provides a brief analysis of three sample notice requirement provisions contained within differing site conditions clauses. The three provisions discussed are the 1976 edition of American Institute of Architects (AIA) Document A201, the 1987 edition of that document, and the General Services Administration (GSA) Standard Form 23A, 1969 edition. This discussion is followed by a comparison of the three provisions of those sample contract forms. The balance of this chapter discusses a sampling of cases in which the notice given to the owner was deemed to be sufficient, notwithstanding some technical or procedural defect, or in which the notice was found to have been waived by some conduct of the owner. Finally, there is a review of those cases in which the contractor did not prevail in its claim for an equitable adjustment for a differing site condition because of failure to follow the notice provisions contained in the differing site conditions clause.

AIA Document A201–1976 edition notice requirement. The 1976 edition of AIA Document A201 permitted an equitable adjustment, by change order, for a differing site condition claim:

upon claim by either party made within twenty days after the first observance of the conditions. [§ 12.2.1]

This section must be read in conjunction with the section that follows, pertaining to claims for additional costs:

If the Contractor wishes to make a claim for an increase in the Contract Sum, he shall give the Architect written notice thereof within twenty days after the occurrence of the event giving rise to such claim. This notice shall be given by the Contractor before proceeding to execute the Work, except in an emergency endangering life or property in which case the Contractor shall proceed in accordance with Paragraph 10.3. [§ 12.3.1]

If not read with § 12.3.1, § 12.2.1 could convey the misconception that the contractor could proceed with work, after having discovered a differing site condition, without obtaining written approval or without submitting a written change order request. Section 12.3.1 should therefore be read to complement Section 12.2.1. To obtain additional compensation in connection with a differing site condition, it is prudent for the contractor to not proceed with work before submitting a written claim for additional costs. In either case, the literal reading of these sections requires only that notice be given by the contractor before proceeding with work; neither clause actually requires, on its face, *approval* of the change order request. Similarly, neither clause actually requires an agreement that there is a differing site condition for the contractor to proceed with work and obtain additional compensation later. In other words, the literal reading of these two sections simply requires *notice,* not approval by the owner. Indeed, as discussed later, the cases support such a literal reading and effectively favor the contractor in connection with a request for an equitable adjustment whenever it has given some type of notice involving a differing site condition.

Unlike § 12.2.1, § 12.3.1 requires *written* notice, and the contractor should give this notice before proceeding with construction performance. Although it appears that the contractor should only comply with notice requirements under § 12.2.1 when changed conditions are encountered, some courts have interpreted the 1976 edition of AIA Document A201 to also require notification under § 12.3.1. Such was the case in North Harris County Junior College District v. Fleetwood Construction Co.,[1] which held that the contractor did not have to comply with the notice requirements of § 12.3.1 only because the owner failed to acknowledge the concealed conditions on which it had received notification under § 12.2.1. As a result, "[c]ollege (the owner) breached the contract and thereby relinquished its procedural rights concerning change orders and claims for additional costs when it refused to acknowledge the 'concealed conditions'."[2] Another case, discussed in detail in **§ 3.3**, held that the notice requirements of the change order provision are *not* applicable to the differing site conditions clause.[3]

AIA Document A201–1987 edition notice requirement. Similar to the 1976 edition notice requirement is that contained in the 1987 edition on discovery of a "concealed or unknown condition." The 1987 edition requires that:

> notice by the observing party shall be given to the other party promptly before conditions are disturbed and in no event later than twenty one days after first observance of the conditions. [§ 4.3.6]

---

[1] 604 S.W.2d 247 (Tex. Civ. App. 1980) writ ref'd n.r.e.

[2] *Id.* at 254.

[3] Olshan Demolishing Co. v. Angleton Indep. Sch. Dist., 684 S.W.2d 179 (Tex. App. 1984) writ ref'd n.r.e.

In addition to providing an extra day within which to give notice, the only difference in these two provisions is the nonsubstantive change of "before conditions are disturbed" instead of the previous language, "before proceeding to execute the work." As a practical matter, the new edition of the AIA document more specifically discourages the contractor from proceeding with the work until the architect has actually investigated the condition and made a determination about the extent to which it differs or was concealed.

Although the 1987 edition contains no separate requirement that the notice of the concealed or unknown condition be in writing, § 4.3.7 requires written notice before work if the contractor wishes to claim increased compensation.

The notice requirements of AIA Document A201, 1987 edition, for differing site conditions and related matters are summarized in **Table 3–1**.

General Services Administration Standard Form 23A–1969 edition. The GSA contract is used in projects involving federal governmental agencies. Its provision dealing with differing site conditions is in paragraph four, which requires that:

Table 3–1

Notice Requirements in AIA Document A201–1987 Edition

	§ 4.3.6	§ 4.3.7	§ 4.3.8	§ 4.3.9
Subject Matter	Concealed or unknown conditions	Claims for additional cost	Claims for additional time	Injury or damage to person or property
Notice Requirements	Notice given promptly but no later than 21 days after first observance	Written notice before proceeding with work	Written notice before proceeding with work	Written notice within a reasonable time not exceeding 21 days after first observance
Information Contained in Notice	Information concerning conditions observed	Details concerning reason for additional costs	Estimate of cost and probable effect of delay	Sufficient detail to enable other party to investigate matter (must still comply with §§ 4.3.7 and 4.3.8 if claim for additional time or costs involved)

Table 3–2

Comparison of Differing Site Conditions Clauses

	AIA Document A201–1976 Edition	AIA Document A201–1987 Edition	GSA Standard Form 23A
Notice Requirements	Give notice and claim within 20 days after first observance of changed condition (or "occurrence" in the case of the actual claim)	Give notice and claim within 21 days after first observance of changed condition (or "occurrence" in the case of the actual claim)	Give notice promptly (unless government extends time)
Type of Notice	Notice must be in writing if the contractor wishes to increase the contract sum	Notice must be in writing if the contractor wishes to increase the contract sum	Notice must be in writing if the contractor wishes to increase the contract sum
Conditions Covered	Type I—conditions at "a variance" with those in contract documents	Type I—conditions that "differ materially" from those in contract documents	Type I—conditions that "differ materially" from those in contract documents
	Type II—conditions that "differ materially" from conditions ordinarily encountered or inherent in work of same type on contract	Type II—conditions that "differ materially" from conditions ordinarily encountered or inherent in work of same type in contract	Type II—conditions that "differ materially" from conditions ordinarily encountered or inherent in work of same type in contract
Remedy	Contract sum equitably adjusted by change order	Architect will cause an increase or decrease in costs or time by recommending equitable adjustment	Contracting officer will make equitable adjustment
Action by Contractor Once Conditions Detected	"Notice shall be given before proceeding to execute the work"	Notice must be given "before conditions are disturbed"	Notice must be given "before conditions are disturbed"

The Contractor shall promptly, and before such conditions are disturbed, notify the Contracting Officer in writing of: (1) subsurface or latent physical conditions at the site differing materially from those indicated in this contract, or (2) unknown physical conditions at the site, of an unusual nature, differing materially from those ordinarily encountered and generally recognized as inhering in work of the character provided for in this contract.

The GSA clause specifically delineates the procedure to be followed by the contractor when an unexpected or differing site condition is encountered. It differs minimally from the AIA differing site notice clause in its requirement for a written notice. Unlike the AIA forms, the notice is required to be given promptly, and not within a specific number of days. Nevertheless, as a practical matter, it would seem that in either the case of a requirement of prompt notice or notice within 21 days, the contractor should not proceed with work until the matter has been brought to the attention of the on-site engineer and not before the on-site engineer has had the opportunity to inspect the condition personally. Ideally, a written change order would be obtained before proceeding further with the work, but the practical reality is such that this does not usually occur.

A summary comparison of the differing site conditions clauses discussed is in **Table 3–2**.

§ 3.2 Comparison of Differing Site Conditions Clauses

Although the discussion so far has focused on the importance of making a written claim for a differing site condition when the contractor seeks an increase in the contract amount, that is not the only reason why the submission of a claim in writing is the prudent course of action. It is equally important to file a written claim when a differing site condition is encountered if the contractor plans to seek an addition to the amount of time allowed for completion of the work. Both the 1976 and 1987 editions of AIA Document A201 require a written change order when the contractor will seek an extension to the contract time. (See § 12.1.1 of the 1976 edition and § 4.3.8 of the 1987 edition.) Similarly, GSA Standard Form 23A, 1969 edition, requires the written notice with respect both to an increase in the contract price and a change in the time required for performance of the contract (§ 4(a)).

§ 3.3 Examples of Sufficient Notice

In most cases involving the recoverability of an equitable adjustment for a differing site condition, in which the issue turns on the timeliness or general sufficiency of the notice, courts will often make the award, although the analysis appears results oriented. For this reason, it is difficult to unequivocally state a rule on which the contractor may universally rely. Rather, the results of a particular case may vary with the equities of that case, its peculiar facts, the notice provision of the contract involved, as well as the jurisdiction of the case. Perhaps because the claim for additional compensation is theoretically based on an "equitable adjustment," and because this is the language frequently encountered in a differing site conditions clause, when courts do find in favor of a contractor's claim for an equitable adjustment for a differing site condition, the finding is based to some extent on waiver, estoppel, lack of prejudice to the owner, or the like. Sometimes, even when waiver, estoppel, lack of prejudice, or some similar equity is not found, actual notice of the differing site condition on the part of the owner is found to be sufficient to overcome the owner's defense that the notice was deficient. In other words, actual notice or knowledge on the part of the owner will sometimes obviate the need for a written notice as required by a literal reading of the contract in question.[4] On the other hand, even though numerous jurisdictions excuse noncompliance with the notice provision in question on the basis of waiver or estoppel, some courts have held that the contractor must show the notice provision to be inapplicable and may also require "clear, decisive, and unequivocal conduct on the part of an authorized representative of the [owner] indicating that it would not insist on adherence to the agreement."[5] In another case, the court held that specificity in the original notice of the differing site condition was not required; rather, the mere indication of a changed condition was sufficient.[6]

What can be generalized about virtually all of these cases is that, even when technical or procedural adherence to a differing site conditions notice provision is not required, two things are required: first, the contractor must provide some type of notice to the owner of the differing site condition, or else the owner must otherwise obtain actual knowledge of the differing site condition; and second, the contractor must file a written claim for an increase in the contract price in accordance with the procedures for a change order. Said differently, although the claim for additional compensation almost uniformly must be in writing, to support a change order for a differing site condition, the contractor often only has to provide notice to the owner of the differing site condition claim, at a point close in time to its discovery, to support its later written change order request for that condition.

In *Lindbrook Construction, Inc. v. Mukilteo School District,*[7] waiver by the owner overcame the failure of the contractor to meet the written notice requirement of the differing site conditions clause. There, a general contractor entered

---

[4] *See, e.g.,* Frederick Snare Corp. v. Maine-New Hampshire Interstate Bridge Auth., 41 F. Supp. 638 (D.N.H. 1941).

[5] Glynn v. City of Gloucester, 401 N.E.2d 886, 892 (Mass. 1980) (involving governmental project and statutorily required notice provision).

[6] S.J. Groves & Sons Co. v. State, 273 S.E.2d 465 (N.C. Ct. App. 1980).

[7] 76 Wash. 2d 539, 458 P.2d 1 (1969).

into a contract with a school district to perform grading and drainage work. The contract documents represented that the fill material required would be available from the areas of proposed cuts. The construction company experienced an extensive shortage of fill material, requiring it to open and use a borrow pit operation. Additionally, the company was required to perform extra excavation in several areas of the project.

The notice provision of the contract provided:

> If the Contractor claims that any instructions by drawings or otherwise involve extra cost under this contract, he shall give the Architect written notice thereof within a reasonable time after the receipt of such instructions, and in any event before proceeding to execute the work, and the procedure shall then be as provided for changes in the work. No such claim shall be valid unless so made.[8]

The trial court held that the written notice requirement had been waived because the architect knew of the work, directed that it proceed, and indicated that the work was not an extra. This judgment was affirmed on appeal, over a strong dissent, despite the school district's contention that there was no testimony or evidence that the school directors knew that the extra work was being done or that the school district was aware of the dispute, the general contractor's claim, or of any potential liability beyond its written contract.

The Washington Supreme Court held in *C.W. Bignold v. King County*[9] that failure to give written notice would not bar a contractor's claim based on quantum meruit. That case involved a unit price contractor in which the county orally ordered the contractor to proceed with work after discovering a differing site condition. The contract was for the construction of a secondary county road. The contractor discovered that a substantial portion of the excavated material included huge unsuitable boulders and was too wet to be used for embankment purposes. This made necessary the contractor's decision to open new additional borrow pits at distances greater than those originally contemplated. The owner contended, among other things, that § 4.04(A) of the specifications required notice in writing of subsurface or latent physical conditions differing materially from those indicated in the contract. The findings of the trial court indicated that the contractor gave timely notice of the differing site conditions and that the owner immediately became aware of the conditions once they developed and ordered the contractor to perform the changes and extra work. Relying on *Frederick Snare Corp. v. Maine-New Hampshire Interstate Bridge Authority,*[10] the appellate court held that "[u]nder such conditions, the [owner] cannot defeat recovery by a contractor even if no written notice was given."[11]

---

[8] 458 P.2d at 3.

[9] 65 Wash. 2d 817, 399 P.2d 611 (1965).

[10] 41 F. Supp. 638 (D.N.H. 1941).

[11] C.W. Bignold v. King County, 65 Wash. 2d 817, 399 P.2d 611, 615 (1965).

In *Davidson & Jones, Inc. v. North Carolina Department of Administration,*[12] the court ruled for the contractor by finding that the differing site conditions notice requirement did not prevent the contractor from filing his claim for compensation, as distinguished from his notice, until after the completion of the work. The general contractor entered into a contract with the North Carolina Department of Administration and the University of North Carolina to build an addition to the library on the University's Chapel Hill campus. The contractor encountered a massive overrun (exceeding 400 percent) due to the amount of rock to be excavated under the contract and requested compensation for the extra excavation and certain duration-related costs. The bid documents, made part of the contract, instructed the bidder to include 800 yards of rock excavation in its base bid and requested a unit price for computing adjustments to the contract for rock above and below this quantity. The general contractor's proposal contemplated eight weeks of rock removal based on the 800-cubic-yard quantity. By the time the owner responded to the general contractor's request for additional payment, the general contractor was seeking payment for both rock excavation at the unit price and duration-related costs it expected to incur after the regularly scheduled completion date. Although the owner agreed to pay for some of the requested items, it instructed the general contractor to pursue the statutory dispute resolution process for payment on the other items when the project was finished. The project was subsequently completed by the general contractor within the time extensions granted. The notice provision in the contract stated:

> Should the contractor consider that as a result of any instructions given in any form by the Engineer or Architect, he is entitled to extra cost above that stated in the contract, he shall give written notice therefore to the Engineer or Architect without delay, and shall not proceed with the work affected until further advised except in an emergency involving the safety of life or property, which condition is covered in Article 12 and 15. No claim for extra compensation will be considered unless the claim is so made.[13]

After exhausting the required administrative processes, the general contractor filed this lawsuit, and the trial court concluded that it was entitled to payment at the unit price for rock excavated and duration-related expenses. The appellate court reversed the decision as to duration-related expenses because the general contractor had already excavated most of the extra rock by the time it requested extra compensation (rather than giving timely notice and stopping work as contemplated by the notice provision). The supreme court, however, rejected this argument and agreed with the trial court. It reasoned that the extra duration-related costs were not incurred until after the scheduled completion date for the entire project, after the general contractor already had change orders in hand for the removal of the extra rock. Additionally, the supreme

---

[12] 315 N.C. 144, 337 S.E.2d 463 (1985).

[13] *Id.* at 149–50, 337 S.E.2d at 466.

court, in its analysis of the notice provision, concluded that at the time the duration-related expenses were about to be incurred, there was no "work about to be affected." Hence, the court found that there was no requirement under the notice provisions that the general contractor stop what it was then doing.

Often a contractor will be able to recover additional compensation for a differing site condition because the owner is found to be estopped from defending on the ground of noncompliance with the notice provision. This was the case in *Kenny Construction Co. v. Metropolitan Sanitary District.*[14] In that case, the general contractor agreed to construct a sewer tunnel. Unfavorable soil conditions (water, running sand, and silt, among others) were encountered, resulting in a modification of the contract as to the contractor's construction methods. The contractor's claim for additional compensation for extra labor was submitted to the owner upon the completion of the entire tunnel and denied. The changed condition clause of the contract read, in pertinent part, as follows:

> Should the contractor encounter during the progress of the work subsurface conditions at the site materially differing from any shown on the contract drawings or indicated in the specifications or such conditions as could not reasonably have been anticipated by either the Metropolitan Sanitary District of Greater Chicago or the contractor, which conditions will materially affect the cost of the work to be done under the contract, the attention of the Chief Engineer must be immediately called to such conditions before they are disturbed. The Chief Engineer shall thereupon promptly investigate the conditions, and if he finds that they do so materially differ, the contract may with his written approval, be modified to provide for increase or decrease of cost and/or differences in time resulting from such conditions. Any increase in costs resulting therefrom shall be subject to approval by the Board of Trustees.[15]

The trial court found for the contractor and awarded it the damages for completing the south end of the tunnel. The appellate court reversed, however, and held that the changed conditions clause was part of the "extra work" provisions and therefore subject to the provision requiring, among other things, that the claim must have been presented to the chief engineer within 30 days following the end of the month in which the extra work was performed.

In reversing the appellate court, the Illinois Supreme Court held that the changed conditions provision stood independently from the "extra work" provisions, all as urged by the general contractor, because those provisions relate to different problems. Although the chief engineer did not provide a written approval for the modification of the contract, and there had been no written approval for a cost increase by the board of trustees, the chief engineer had told the contractor to proceed with the work, with payment to be made at a later date because it was more realistic to evaluate the work after it had been

---

[14] 52 Ill. 2d 187, 288 N.E.2d 1 (1971).

[15] 288 N.E.2d at 3.

completed. For this and other reasons, the supreme court held that the owner was estopped from seeking to avoid liability on the grounds that the change was not approved in writing.

As noted, at least one court has specifically held that the notice requirements of the change order provisions are *not* applicable to the differing site conditions clause.[16] There, the contractor agreed to demolish a two-story building on a school district campus. The contract included the demolition and removal of the building, foundation, paving, and sidewalks. During the work, the contractor discovered three additional slabs underneath the building that were not shown on the architectural drawings upon which the contractor had based its bid. The contractor notified the school district's architect of the condition that had been encountered and requested additional payments for the work that would be required to remove the slabs. The school district rejected the request for additional payment. The three slabs were removed before the contractor filed a written request for additional compensation.

The school district argued that because the contractor failed to give written notice (as required by paragraph 12.3.1 of the general conditions of the contract) before executing the work, the contractor's claim was invalid. The contractor argued that this provision was inapplicable and urged that paragraph 12.2.1 (concealed conditions) was controlling, stating that the contract sum would be equitably adjusted by change order if either party made a claim within 20 days after the first observance of the concealed conditions.

Reasoning that the parties intended for concealed conditions to be treated differently from claims for additional costs, the court of appeals reversed the trial court in favor of the contractor, holding that the contractor was not required to follow the notice requirements of paragraph 12.3.1.

A similar result was reached by the court of claims in *Shepherd v. United States,*[17] which ruled for the contractor by finding a lack of prejudice to the government in the deficient notice. There the general contractor agreed to construct the Cumberland Oil Field protective levies, which were a part of the Dennison Dam and Reservoir Project on the Red River between Oklahoma and Texas. The contractor encountered and advised the contracting officer of wet material differing from that reflected in the drawings and specifications. It was implied that this notice was given orally.[18] This caused the contractor to incur increased costs in excavating, hauling, and placing the material. At the time, however, the contractor did not state that this was an unforeseen condition entitling it to an equitable adjustment. Rather, the contractor sought permission to waste the wet material and to obtain other material for the fill from borrow pits, which would have greatly increased its compensation. Because the contractor did not claim an

---

[16] Olshan Demolishing Co. v. Angleton Indep. Sch. Dist., 684 S.W.2d 179 (Tex. App. 1984) writ ref'd n.r.e.

[17] 113 F. Supp. 648 (Ct. Cl. 1953).

[18] *Id.* at 651.

unforeseen condition, the contracting officer apparently thought that he was not required to make an investigation to determine whether or not the different site condition had actually been encountered, warranting an increase in the contract price. The contract read in pertinent part:

> the attention of the contracting officer shall be called immediately to such conditions before they are disturbed.[19]

Based on this clause and the above facts, the court determined that the contractor had done all the contract required; namely, the contractor called the contracting officer's attention to the conditions before they were disturbed. The court reasoned that the owner was not prejudiced by the contractor's failure to claim that the conditions differed from those that it had a right to expect because the contract imposed the duty on the contracting officer to determine whether the conditions were materially different, even if the contracting officer discovered the conditions.

Relying on *Shepherd,* the claims court found that oral notice was sufficient in *General Casualty Co. of America v. United States,*[20] a case involving an airport construction project. In this case, the contractor encountered several thousand cubic yards of hard shell (which was slowing operations) not shown by the drawings, specifications, or borings, but the contractor did not claim a changed condition. The contractor protested orally to the resident engineer in September 1943, and a month later in a letter to the engineer in charge of design. Previously, the contractor's engineer and two successive superintendents also made oral representations about the changed conditions to the government's resident engineer and his assistants. The contract stated, in pertinent part:

> Should the contractor encounter, or the Government discover, during the progress of the work subsurface and/or latent conditions at the site materially differing from those shown on the drawings or indicated in the specifications, or unknown conditions of an unusual nature differing materially from those ordinarily encountered and generally recognized as inhering in the work of the character provided for in the plans and specifications, the attention of the contracting officer shall be called immediately to such conditions before they are disturbed. The contracting officer shall thereupon promptly investigate the conditions, and if he finds that they do so materially differ, the contract shall, with written approval of the head of the department or his duly authorized representative, be modified to provide for any increase or decrease of cost and/or difference in time resulting from such conditions.[21]

Holding, as in the *Shepherd* case, that the resident engineer was the authorized representative of the contracting officer, the court held that the contractor

---

[19] *Id.*

[20] 127 F. Supp. 805 (Ct. Cl. 1955).

[21] *Id.* at 811.

did all that it was required to do under the contract when it gave notice to the resident engineer at the site.[22] The court said it would be "inane to suppose that the resident engineer was at the site for no purpose."[23]

Finding the owner was estopped to avoid increased compensation to the contractor in reliance on procedural technicalities, the court in *Moorhead Construction Co. v. City of Grand Forks*[24] held for the contractor. The general contractor sought to recover from the owner extra expenses and lost profits incurred in building a sewage treatment facility. The project was divided into two phases to be performed by separate contractors. Phase 1 included primarily the earth work and site preparation for four aerated-anaerobic treatment ponds, including installation of piping and appurtenances such as foundations for compressor and meter buildings. Phase 2 consisted of completing the buildings, constructing manhole installations and access bridges into and over the ponds, and installing all electrical and mechanical equipment. During the construction of Phase 2, the contractor encountered unstable soil conditions in the cell bottoms and on the embankments, which forced it to work by different, more expensive methods, without heavy equipment. This was because the Phase 1 contractor failed to obtain 90-percent compaction, an implied warranty in the Phase 2 contract. The contract provided, in pertinent part, as follows:

> Should the Contractor encounter or the Owner discover during the progress of the work subsurface or latent physical conditions at the site, differing materially from those indicated in this contract, or unknown physical conditions at the site of an unusual nature, differing materially from those ordinarily encountered and generally recognized as inherent in the work of the character provided for in the contract, the Engineer shall be promptly notified in writing of such conditions before they are disturbed. The Engineer will thereupon promptly investigate the conditions and if he finds they do so materially differ and cause an increase or decrease in the cost of, or the time required for the performance of the contract, an equitable adjustment will be made and the contract modified in writing accordingly.[25]

The city contended that the contractor was precluded from employing the changed conditions clause because it failed to carry the administrative process to a final written contract modification. The appellate court, however, affirmed the trial court's conclusion that the city waived its rights under the contract and was estopped from denying the general contractor an equitable adjustment. This was based on the district court's findings to the effect that the owner's consulting engineer, as agent for the owner, determined that a changed condition did in fact exist, and it was the owner, not the general contractor, that was responsible

---

[22] *Id.* at 812.

[23] *Id.*

[24] 508 F.2d 1008 (8th Cir. 1975).

[25] *Id.* at 1013 n.4.

for the failure to carry the administrative process to a final written contract modification.

Oral notice has been held to suffice, as support for a claim for additional compensation, especially when it is combined with evidence that the owner orally directed or allowed the work to proceed.[26] In that case, the court held that oral notice to the engineer, who allowed work to proceed, was sufficient to allow recovery for differing site conditions costs. The general contractor encountered rock, subsurface water, and unstable soil, for which it allegedly incurred additional expense during excavation for the installation of underground fuel storage tanks. The trial court awarded the contractor damages for the extra work done, prejudgment interest, and its attorneys' fees and court costs. The owner appealed, claiming that the general contractor was not entitled to additional compensation due to its failure to notify the owner of abnormal conditions and failure to obtain an executed change order before commencing work. In the owner's response to the general contractor's bid qualification, the owner stated that:

> Since rock, unstable soil or water are not included in any surveys we might provide, these unknown subsurface conditions are not covered by this General Condition. In this case, we do not know what normal subsurface conditions may exist and have not asked you to verify what these conditions might be or to assume responsibility for same. Your bid is assumed to reflect "normal" excavation: if abnormal conditions are encountered, we are to be notified so that we may verify this condition and reach an agreement with you on the manner in which to proceed.[27]

The owner argued that there was insufficient evidence to support the jury's finding that the contractor performed the extra work upon the prior approval of the owner's engineer. The appellate court overruled this point with respect to the rock excavation and sustained it with respect to the retaining wall and removal of water. In finding that there was sufficient evidence to support the jury's finding with respect to the rock excavation, the court relied on the testimony of the general contractor and the owner's engineer to the effect that the engineer was notified when the rock was discovered, he was informed that this excavation would cost extra, and he allowed this work to take place.

D. Federico Co. v. New Bedford Redevelopment Authority[28] dealt with a statutorily prescribed notice provision similar to the one contained in the GSA provision. The general contractor brought its action against a redevelopment authority seeking extra excavation costs incurred in the construction of two earth-filled bulkhead piers. Although the trial court found that the contractor failed to

---

[26] City of Fort Worth v. Gene Hill Equip. Co., 761 S.W.2d 816 (Tex. App.—Fort Worth 1988, no writ).

[27] *Id.* at 818.

[28] 723 F.2d 122 (1st Cir. 1983).

adequately perform the required pre-bid investigation, it also found that the owner possessed information, not provided to the bidders, describing the obstructions that were eventually encountered. The bid documents provided, in pertinent part:

> Any discrepancies which may be discovered between actual conditions and those represented by the drawings and maps shall at once be reported . . . .
>
> If on the basis of the available evidence the Local Public Agency determines that an adjustment of the Contract price and/or Time is justified, the procedure shall then be provided in Section—CHANGES IN THE WORK under GENERAL CONDITIONS, PART 1.[29]

General Laws of the Commonwealth of Massachusetts, Chapter 30, § 39N, similarly stated:

> Construction Contracts; equitable adjustment in the contract price for differing subsurface or latent physical conditions. Every contract . . . shall contain the following paragraph in its entirety and an awarding authority may adopt reasonable rules and regulations in conformity with that paragraph concerning the filing, investigation and settlement of such claims.
>
> If, during the progress of the work, the contractor or the awarding authority discovers that the actual subsurface or latent physical conditions encountered at the site differ substantially or materially from those shown on the plans or indicated in the contract documents either the contractor or the contracting authority may request an equitable adjustment in the contract price of the contract applying to work affected by the differing site conditions. A request for such an adjustment shall be made in writing and shall be delivered by the party making such claim to the other party as soon as possible after such conditions are discovered. Upon receipt of such claim from a contractor, or upon its own initiative, the contracting authority shall make an investigation of such physical conditions, and, if they differ substantially or materially from those shown on the plans or indicated in the contract documents or from those ordinarily encountered and generally recognized as inherent in work of the character provided for in the plans and contract documents and are of such a nature as to cause an increase or decrease in the cost of performance of the work or a change in the construction methods required for the performance of the work which results in an increase or decrease in the cost of the work, the contracting authority shall make an equitable adjustment in the contract price and the contract shall be modified in writing accordingly.[30]

After the underwater obstructions were encountered in September 1975, the contractor notified the engineers employed by the owner, and the owner and the contractor were in constant communication until June 1976, to determine the identity, location, and character of the obstructions. Even the meeting minutes of the owner reflected its awareness and involvement in discussions

---

[29] *Id.* at 126 n.3.

[30] *Id.*

concerning the problem. Once the contractor determined the exact character of the obstructions, it informed the owner, in writing, that the problem was beyond the scope of the contract and requested further instructions as to the manner in which the owner desired to proceed. Based on these facts, the court held that the contractor's notice could not be characterized as inadequate.

When the owner denies a claim arising under a differing site conditions clause, based on the merits of the claim, the notice provisions may be deemed waived.[31] In that case, a general contractor agreed to build a major water delivery pipeline for the defendant municipality. In performing excavation in the area of the project known as Lambert's Meadow, the contractor encountered an extraordinary concentration of boulders, necessitating additional work. Before such conditions were disturbed, the contract required that the general contractor give written notice to the engineer of the unknown physical condition encountered. At trial, the court found that the contractor did not comply with the requirement of prompt written notice and therefore did not determine whether there actually was a changed condition. The court of appeals reversed that part of the trial court's decision, holding that the owner waived the contractor's failure to provide notice because it denied the contractor's claim on the merits.

Mere actual notice is often sufficient to meet even a written notice requirement. The court in *Frederick Snare Corp. v. Maine-New Hampshire Interstate Bridge Authority*[32] held that the written notice defense was merely "technical" and would not bar recovery for a differing site condition claim of which the owner had actual notice.

There, a contractor had to perform additional excavation required in its construction of a bridge when it discovered that the bidding plan did not represent the true, underwater surface conditions as to the location of ledge or rock foundation. The contract provided that:

> Should the contractor encounter subsurface and/or latent conditions at the site materially differing from those shown on the Plans or indicated in the Specifications, he shall immediately give notice to the Engineer of such conditions, before they are disturbed. The Engineer shall thereupon promptly investigate the conditions, and if he finds that they materially differ from those shown on the Plans or indicated in the Specifications, he shall at once make such changes in the Plans and/or Specifications as he may find necessary, any increase or decrease of costs resulting from such changes to be adjusted in the manner provided herein for adjustments as to extra and/or additional work and charges.[33]

The contract further provided that notice meant written notice. Because no such written notice was provided, the defense contended that the contractor was

---

[31] Metropolitan Paving Co. v. City of Aurora, 449 F.2d 177 (10th Cir. 1971).

[32] 41 F. Supp. 638 (D.N.H. 1941).

[33] *Id.* at 645.

barred from recovering the costs of the extra work performed. The court held that this position was just "too technical to work out justice between the parties."[34] Reasoning that notice is for the protection of the parties, and that writing only furnishes proof of such notice, the court rejected this defense, finding that the defendant had actual notice and the parties were familiar with all facts requiring notice.

Also ruling against the owner on the "technical" defense was *Metropolitan Sewerage Commission v. R.W. Construction, Inc.*[35] The general contractor agreed to build a contract for the construction of a gravity flow sewage system. The general contractor sought an equitable adjustment and compensation, both because of the unanticipated artesian water that it encountered and because the use of compressed air was not feasible. The contract provided, in pertinent part:

> Should the Contractor encounter, or the Sewage Commission discover during the progress of the work, subsurface and (or) latent conditions at the site materially differing from those shown on the drawings or indicated in the specifications, the attention of the Engineer shall be called immediately to such conditions before they are disturbed. The Engineer shall thereupon promptly investigate the condition and, if he finds that they materially differ from those shown on the drawings or indicated in the specifications, he shall at once, make such changes in the drawings and (or) specifications as he may find necessary and any increase or decrease of and (or) difference in time resulting from such changes shall be adjusted.[36]

The court denied the owner's contention that the contractor could not take advantage of the changed conditions clause of the contract because it did not notify the engineer of the conditions encountered, holding that the clause did not require a formal notice of claim, and it was undisputed that the owner, through its engineer, simultaneously discovered the same changed conditions as did the contractor.

It is not generally necessary that a differing site condition notice be as detailed as the actual claim for extra compensation that will most often be filed later, even after the work is performed. What is necessary is notice of the general nature of the condition and, preferably, notice that an equitable adjustment will be sought on account of the condition. *S.J. Groves & Sons Co. v. State,*[37] which dealt with the sufficiency of the notice of the differing site condition, involved the defense that it failed to be detailed enough. In *Groves,* the contractor argued that excessive moisture and unstable soil were such changed conditions at the work site so as to entitle it to an equitable adjustment in compensation. The contract provided in pertinent part as follows:

---

[34] *Id.*

[35] 72 Wis. 2d 365, 241 N.W.2d 371 (1976).

[36] 241 N.W.2d at 374–75.

[37] 273 S.E.2d 465 (N.C. Ct. App. 1980).

> Should the Contractor encounter or the Commission discover during the progress of the work conditions at the site differing materially from those indicated in the contract, which conditions could not have been discovered by reasonable examination of the site, the Engineer shall be promptly notified in writing of such conditions before they are disturbed. The Engineer will thereupon promptly investigate the conditions and if he finds they do so materially differ and cause a material increase or decrease in the cost of performance of the contract, an equitable adjustment will be made and a supplemental agreement entered into accordingly.[38]

On appeal, the owner argued that the contractor's notice was insufficient because it failed to spell out in detail the exact nature and extent of the unclassified excavation work it was claiming under a changed condition. The appellate court affirmed the trial court's finding as to the sufficiency of the notice, stating "[a]t this point in its claim, plaintiff was not required to itemize the fine points and particulars which subsequently would be necessary in the proof of its claim. All that was necessary at this junction was a 'forceful indication of changed conditions and demand for equitable compensation.'"[39]

If a set of principles can be extrapolated from the preceding discussion, it is that, more frequently than not, equitable considerations prevail. Although the extent to which traditional equity principles discussed in the cases vary from case to case, it is readily apparent that mere technical or "legalistic" defenses to a claim for an equitable adjustment for a differing site condition will not normally result in the denial of the equitable adjustment. The final outcome is always particular to the case in question, the jurisdiction involved, the equities involved, and the particular contract language involving the differing site condition notification requirements. However, as these cases demonstrate, when there is some attempt at notification to the owner, with knowledge on the part of the owner of the differing site condition, a technical deficiency in the delivery or content of the notification is not usually a bar to recovery. This will more likely be the case in those situations in which, as discussed, the owner has directed or at least allowed the work to proceed after being notified of the differing site condition or after having independently come into knowledge of the condition. It is in those instances when language of waiver, estoppel, and lack of prejudice are seen.

Of course, even though the equities of the particular case may run in favor of the contractor, this is an after-the-fact analysis. The preventive and conservative approach to be taken whenever possible is to prepare and submit written notice of the differing site condition in as much specificity as possible at the time. This should be followed as quickly as possible with an actual detailed claim for an equitable adjustment, sufficient for the purposes under the contract documents in question. As discussed in **§ 3.4**, the failure to follow this conservative

---

[38] *Id.* at 471.

[39] *Id.* at 493.

approach can result in a complete bar to the contractor's right of recovery for its claim for an equitable adjustment based on a differing site condition.

§ 3.4 Cases in Which Contractor Did Not Prevail

In some cases, courts have ruled against contractors when the only notice given was oral, even when followed by a written claim for additional compensation at the conclusion of the work. Although defending against a claim based on the notice being only oral seems, more often than not, an inadequate defense, it will occasionally prevail. From the contractor's point of view, the sensible and conservative action to take is to provide written notice of a differing site condition.

In *Blankenship Construction Co. v. North Carolina State Highway Commission,*[40] the general contractor agreed to perform certain road construction for the North Carolina State Highway Commission. Although the contractor's site investigation and the owner's subsurface summary reflected only 135,000 cubic yards of rock to be excavated, the contractor actually encountered between 750,000 and 800,000 cubic yards of rock on the job. The concealed conditions clause provided, in pertinent part, that:

> Should the Contractor encounter or the Commission discover during the progress of the work conditions at the site differing materially from those indicated in the contract, which conditions could not have been discovered by reasonable examination of the site, the engineer should be promptly notified in writing of such conditions before they are disturbed. The Engineer will thereupon promptly investigate the conditions and if he finds they so materially differ and cause a material increase or decrease in the cost of performance of the contract, an equitable adjustment will be made and a supplemental agreement entered into accordingly.[41]

After discovering the unexpected rock conditions, the contractor orally notified the engineer who responded that the work bid was classified so as to preclude a claim for additional compensation under the concealed conditions clause. The contractor proceeded with the work and filed a claim for additional compensation after its completion. The contractor's oral notice was viewed as amounting to a tentative inquiry rather that a forceful indication of changed conditions and a demand for equitable compensation; the appellate court therefore affirmed the trial court's finding that the notice requirement of the concealed conditions clause was not satisfied.

This result has obtained even when it was shown that the owner had actual notice of the differing site condition,[42] although the result may well have been

---

[40] 28 N.C. App. 593, 222 S.E.2d 452 (1976).

[41] 222 S.E.2d at 458.

[42] Inland Bridge Co. v. North Carolina State Highway Comm'n, 30 N.C. App. 535, 227 S.E.2d 648 (1976).

more favorable to the contractor had its case been submitted differently. The case involved a general contractor that agreed to relocate a road. The contractor, in bidding the project, relied, and was entitled to rely, on representations that the moisture content of the field dirt was 40 percent. The owner had actual knowledge that the moisture content, in many instances, greatly exceeded 40 percent and that the soil could not be practically moved and dried. The soil was classified as "unclassified" rather than "unsuitable" by the owner, without alluding to its known condition. The contract provided, in pertinent part, as follows:

> Should the Contractor encounter or the Commission discover during the progress of the work conditions at the site differing materially from those indicated in the contract, which conditions could not have been discovered by reasonable examination of the site, the Engineer shall be promptly notified in writing of such conditions before they are disturbed. The Engineer will thereupon promptly investigate the conditions and if he finds they do so materially differ and cause a material increase or decrease in the cost of performance of the contract, an equitable adjustment will be made and a supplemental agreement entered into accordingly.[43]

The appellate court refused to consider the claim on procedural grounds, ruling that had the contractor desired to sue under the changed conditions provision of the contract, it was necessary to elect to do so before the original trial. The court also found that the contractor failed to comply with the notice-in-writing requirement of the clause and failed to keep cost records as required by the contract. The court therefore concluded that the contractor could not recover under the changed conditions clause in the contract.

When there is neither a written notice of the differing site condition nor any basis for a finding of waiver or estoppel on the part of the owner, the result can be disastrous for the general contractor. This is particularly true if the condition is uncovered early in a project but the claim for equitable adjustment is submitted many months later, at the end of the project.[44] In *Skopek,* a general contractor sought an equitable adjustment in the contract price by reason of its encountering and overcoming actual or latent soil conditions materially different from those indicated in the contract documents. The applicable statute[45] and the contract required a "written request" for such an adjustment. However, the request was not provided until some 16 months after the plaintiff encountered the conditions of which it complained. Finding no evidence that these conditions were ever waived or excused by anyone, the court held that the contractor was not entitled to an equitable adjustment.

Sometimes just a "technical" defense based on insufficient notice will suffice to defeat a claim for equitable adjustment. That was how the owner prevailed in

---

[43] 227 S.E.2d at 650.

[44] Skopek Bros., Inc. v. Webster Hous. Auth., 416 N.E.2d 1006 (Mass. App. Ct. 1981).

[45] See discussion of D. Federico Co. v. New Bedford Redev. Auth., 723 F.2d 122 (1st Cir. 1983), in § **3.3**.

Central Penn Industries, Inc. v. Commonwealth.[46] The general contractor agreed to construct seven miles of a separated highway. Due to a perceived inability to use on-site fill, the contractor incurred additional costs of borrow excavation. The specifications provided as follows:

> Neither the contractor nor the surety shall be entitled to present any claim or claims to the Secretary or the Board of Arbitration of Claims named in the contract, either during the prosecution of the work, or upon completion of the contract, for additional compensation for any work performed which was not covered by the approved drawings, specifications and/or contract, or for any other cause unless he or it shall have given the Secretary due notice in writing of his or its intention to present such claim or claims as hereinafter designated; providing, however, that the contractor or his surety shall not be denied the rights to present any claim which is based on differences and measurements or areas of computations which are not disclosed until preparation of the final certificate.[47]

The written notice was required to be given within 10 days from the inception of the claim. The appellate court reversed the Board of Arbitration Claims' allowance of the contractor's claim for additional compensation because such written notice was not provided. The court held that a letter requesting a change in the specifications, but containing not a hint of a claim, was insufficient to meet this requirement.

In summary, it is important for contractors to follow four points when encountering differing site conditions:

1. Notice should be given in writing of the differing site condition encountered.
2. The notice of the differing site condition should be given to the owner's representative as soon as humanly possible after the discovery of the differing site condition.
3. The notice should contain as much detail as is known about the differing site condition.
4. The notice should include an affirmative indication that a claim will be made for an equitable adjustment to the contract price, an extension to the contract time for completion of the project, or both.

In addition to these four important points, whenever possible, the contractor should also attempt to ensure that the owner's representative is given the opportunity to inspect the differing site condition, and that the representative does, in fact, make such an inspection. Further, the ideal situation will involve an affirmation by the owner's representative of the existence of a differing site

---

[46] 358 A.2d 445 (Pa. Commw. Ct. 1976).

[47] *Id.* at 449.

condition and some agreement as to the nature and extent of that condition. All of these things will ideally occur before proceeding with work that involves the differing site condition. Of course, when the contractor can get advance approval, in the form of a change order, for an equitable adjustment on account of a differing site condition, this should be undertaken. More often than not, however, this is not possible.

When the contractor cannot adhere to these points in connection with making a claim for an equitable adjustment for a differing site condition, the contractor will have to rely to some extent on the equities of the situation to obtain a recovery. The prudent contractor should therefore take heed to document the equities involved to the maximum extent possible. In this fashion, the contractor can effectively deal with its failure or inability to follow or comply with the literal terms of a notice requirement by relying on equitable doctrines such as estoppel, waiver, or lack of prejudice to the owner and the like.

CHAPTER 4

EFFECT OF EXCULPATORY CLAUSES

Frederick S. Cohen
James F. Oliviero

Frederick S. Cohen is the senior partner in the New York City law firm of Ross & Cohen which specializes in construction and government contract law. He is a graduate of Alfred University and Brooklyn Law School in New York. Mr. Cohen is an officer and/or member of the executive committee of the Committee of Construction and Surety Law of the New York State Bar Association, the committee on International Construction Contracts of the International Bar Association, the Public Contract Law Section of the American Bar Association and the Construction Industry Panel of Arbitrators of the American Arbitration Association.

James F. Oliviero is a member of the New York City law firm of Ross & Cohen, which specializes in construction law, representing all facets of the construction industry. A graduate of St. John's Law School in New York, his practice includes litigation involving site condition claims on public and private contracts. Mr. Oliviero is a member of the construction industry panel of arbitration of the American Arbitration Association and the committee on construction and surety law of the New York State Bar Association.

§ 4.1 Introduction

Contractors who bid on construction contracts involving excavation work face the possibility of underbidding the contract in light of the potential for unforeseen and unanticipated subsurface conditions. The types of subsurface conditions are varied. A recent federal government office project in downtown New York City was stopped when excavation uncovered a long-forgotten slave burial ground. Such circumstances put the contractor in a precarious position, torn between the goal of preparing as low a bid as possible, while protecting itself from underbidding the job by including a contingency for any unforeseen or unanticipated subsurface condition arising during excavation. The only instance in which a contractor does not have to balance the risk of unforeseen or unanticipated subsurface conditions with a contingency in its bid estimate is when the contract includes a differing site conditions clause.

Differing site conditions clauses, which are typically found in federal government agency contracts, generally provide that if the contractor can prove that (1) subsurface or latent physical conditions at the site differ materially from those indicated in the contract (Type I condition), or (2) unknown and unusual physical conditions exist at the site that differ materially from those ordinarily encountered and generally recognized as inhering in the work provided for in the contract (Type II condition), the contractor will receive compensation over and above the contract price. Therefore, a contractor can concentrate on submitting the low bid because the contract has a built-in mechanism for extra compensation for unforeseen or unanticipated subsurface conditions.

If the contractor does encounter an unanticipated subsurface condition and the contract does *not* contain a differing site conditions clause, it must then be decided whether the owner or the contractor is responsible for the extra costs associated with the unanticipated subsurface condition. Most construction contracts contain exculpatory clauses whereby the owner attempts to put the risk of unforeseen subsurface conditions on the contractor. Various exculpatory clauses either (1) disclaim liability for any representations of subsurface conditions that prove to be false; (2) advise the contractor that soils information inserted in a bid package is not part of the contract documents, are for informational purposes only, and are not to be relied on by the contractor, and/or (3) impose a duty on the contractor to familiarize itself with the work to be performed and the site conditions. Although exculpatory clauses generally insulate the owner from bearing the costs of unforeseen subsurface conditions, there are certain circumstances that will render an otherwise valid disclaimer ineffective.

The United States Supreme Court has explicitly held that in certain construction contracts there is an implied warranty of the fitness of the plans and specifications. When an owner incorporates *design specifications* in the contract that are to guide the contractor's performance, the owner represents that the contractor will not encounter any unforeseen or unanticipated subsurface conditions.

If the contractor cannot perform its work by following the "design" specifications because it encounters unforeseen subsurface conditions, then the owner will have to bear the extra excavation costs incurred. Therefore, an otherwise valid exculpatory clause disclaiming any representation of subsurface conditions or any liability for unanticipated or unforeseen subsurface conditions will not be upheld by a court.

If the specifications amount to only *performance specifications* (specifications that only set forth the objectives or standards that the contractor is expected to meet), there is no implied warranty of the fitness of the plans and specifications. Performance specifications do not give rise to a representation of subsurface conditions that, if proven false, will render an otherwise valid disclaimer ineffective.

Courts have also held that although general disclaimers may not preclude a contractor's reliance on a positive representation made in a contract, a disclaimer that is closely related to the condition actually encountered will likely bar a contractor's recovery for breach of implied warranty of the fitness of the plans and specifications. An owner will also be liable for the extra costs associated with unanticipated subsurface conditions if it misrepresents the subsurface conditions in the bid or contract documents. Courts have recognized that a misrepresentation of subsurface conditions will render an otherwise valid disclaimer ineffective whether the misrepresentation is innocent, intentional (fraudulent), or negligent. Although the requirements for proving the three types of misrepresentation are slightly different, each requires that the owner make a positive, false representation of subsurface conditions.

Positive representations of subsurface conditions may be made by including information contained in a soils report or in test borings or by not including certain information in the contract that was known only by the owner, could not be reasonably discovered by the contractor, and whereby its exclusion misled the contractor as to the subsurface conditions. A contractor cannot claim that an owner made a false, positive representation of subsurface conditions when the contractor interprets certain information given by the owner, and the contractor's interpretation turns out to be incorrect. Likewise, if the subsurface information given by the owner merely suggests what subsurface conditions might be encountered, a valid disclaimer will shift the extra, unforeseen costs to the contractor.

Whether the contractor claims that a valid disclaimer should be rendered ineffective due to a breach of implied warranty of the fitness of the plans and specifications or some form of misrepresentation, courts generally require that the contractor perform its own reasonable site investigation. The reasonableness of a site investigation depends on the circumstances of each case. Nevertheless, if the contractor could have noticed an unforeseen subsurface condition with a cursory inspection, the court will not render the valid disclaimer ineffective. The balance of this chapter provides a framework for the contractor to analyze these problems and, it is hoped, to correctly estimate their risks.

§ 4.2 Implied Warranty of Plans and Specifications

When a contractor is required to build according to plans and specifications supplied by an owner, the contractor will not be responsible for the expenses incurred as a result of defects in the plans and specifications. This is known as the *Spearin* doctrine,[1] which sets forth the general rule that there is an implied warranty that if specifications are complied with, satisfactory construction will result. In the context of site conditions, the implied warranty is that representations made by the owner regarding these conditions are accurate. Thus, the nature of the warranty is that a contractor has the right to rely on information provided by an owner. A breach of the warranty results when the information reasonably relied on turns out to be false.[2]

An owner may sometimes include in contract documents a disclaimer of any warranty in an attempt to make the contractor responsible for the financial burden caused by an unforeseen condition. The presence of a disclaimer affects whether the contractor had the right to rely on the representations contained in the plans. A general disclaimer will not necessarily preclude a claim for breach of warranty based on positive representations that turn out to be untrue. However, a disclaimer that specifically addresses the particular representation on which a contractor claims to have relied will generally preclude a breach-of-warranty claim, especially in cases in which the representation made was not a positive assertion of existing conditions.

§ 4.3 —Warranty of Design Specifications versus General Disclaimers

The implied warranty of plans and specifications extends only to design specifications when the contractor is explicitly told how the contract is to be performed. A performance specification, on the other hand, sets forth objectives or standards that the contractor is expected to meet, leaving it to the contractor to determine how to achieve those results.[3] The law does not imply a warranty of performance specifications.[4]

A general disclaimer will not insulate an owner from liability for breach of its implied warranty of its plans and specifications. In the *Spearin*[5] case, for

---

[1] United States v. Spearin, 248 U.S. 132, 39 S. Ct. 59 (1918).

[2] Alpert v. Commonwealth, 258 N.E.2d 755 (Mass. 1970).

[3] Utility Contractors, Inc. v. United States, 8 Cl. Ct. 42 (1985), *aff'd,* 709 F.2d 90 (Fed. Cir.), *cert. denied,* 479 U.S. 827, 107 S. Ct. 104 (1986); J.L. Simmons Co. v. United States, 188 Ct. Cl. 684, 412 F.2d 1360 (1969).

[4] Universal Contracting & Brick Pointing Co. v. United States, 19 Cl. Ct. 785 (1990); Stuyvesant Dredging Co. v. United States, 834 F.2d 1576 (Fed. Cir. 1987).

[5] United States v. Spearin, 248 U.S. 132, 39 S. Ct. 59 (1918).

example, a contractor was to build a dry dock based on plans and specifications provided by the government. Pursuant to the plans, the contractor began the project by relocating a section of a six-foot brick sewer that intersected the site and connecting it with an existing seven-foot sewer. When the six-foot sewer broke and the excavation for the dry dock was flooded, it was revealed there was a dam in the seven-foot sewer that had not been shown on the plans. After extensive investigation, the contract was annulled, and the contractor sued to recover its costs and its lost profit. The Court found in favor of the contractor based on the fact that the specifications were very specific concerning the sewer construction, and thus they constituted an implied warranty. Although the contract required the contractor to examine the site, check the plans, and be responsible for the work, the Court held that the implied warranty of the specifications could not be overcome by these general clauses.[6] As the Supreme Court had previously noted:

> We think this positive statement of the specifications must be taken as true and binding upon the Government, and that upon it rather than upon the claimants must fall the loss resulting from such mistaken representations. We think it would be going quite too far to interpret the general language of the other paragraphs [of the Contract] as requiring independent investigation of facts which the specifications furnished by the Government as a basis of the contract left in no doubt. If the Government wished to leave the matter open to the independent investigation of the claimants it might easily have omitted the specifications as to the character of the [soil condition]. In its positive assertion of the nature of this much of the work it made a representation upon which the claimants had a right to rely without an investigation to prove its falsity.[7]

Thus, in a claim based on breach of warranty, an owner's responsibility for the warranty created by its positive representations may not be avoided by inserting general disclaimers into a contract.

§ 4.4 —Specific Disclaimers

Although general disclaimers may not preclude a contractor's reliance on a positive representation made in a contract, a contractor must be aware that a disclaimer that is closely related to the condition actually encountered will likely bar recovery for breach of warranty. A provision in a contract that expressly disclaims the existence of an implied warranty, or that expressly disclaims responsibility for site conditions that are encountered, will preclude a claim for breach of warranty.

---

[6] 248 U.S. at 137.

[7] Hollerbach v. United States, 233 U.S. 165, 172 (1914).

Thus, in *Eastern Tunneling Corp. v. Southgate Sanitation District*,[8] a contractor could not make a breach-of-warranty claim based on inaccuracies in a soils report provided at the time of bidding when the contract included specific disclaimers relating to the accuracy of that report. Because the owner had specifically stated that the information provided was not guaranteed and that the contractor would bear the risk if actual conditions differed,[9] the representations included in the soils report did not constitute implied warranties of their accuracy.

In a similar case,[10] the contractor could not recover the cost of installing dewatering wells even though soil boring logs provided pre-bid had not indicated the presence of a high groundwater table. The court found that there was nothing in the contract to indicate that the boring logs were intended to be relied upon and that, therefore, exculpatory provisions requiring the contractor to assume the risk of adverse subsurface conditions necessitating dewatering made the contractor's reliance on the logs unjustifiable.

An effective disclaimer of an implied warranty was enforced in *D. Federico Co. v. Commonwealth,*[11] in which a contractor had to excavate and provide far more replacement fill than estimated by the contract. The contractor argued that the owner had impliedly warranted that its estimates would be fair approximations of the amounts that would be necessary, and that therefore the contractor was entitled to recover for breach of this warranty. The court, however, rejected this claim based on exculpatory provisions in the contract that specifically precluded such a warranty of the very estimates on which the contractor had claimed to rely.[12]

In *D. Federico,* the court specifically rejected as a basis for recovery the fact that the brief time the contractor had to prepare its bid made it impossible for it to do independent testing.[13] However, other courts have considered this factor and applied it to the contractor's benefit and, as will be discussed in §§ **4.6** through **4.9**, this is an important factor in a contractor's claim based on a misrepresentation rather than a breach-of-warranty theory. For example, in a Montana case, although a contract contained specific disclaimers of its estimates of the amount of borrow the contractor would need, the court held that the evidence in that case could support a finding that the contractor had justifiably relied on the estimate and that the disclaimers would not necessarily preclude such reliance.[14] Although the contractor had inspected the site and inquired about the soils in the area, it had only two to three weeks to submit its bid,

---

[8] 487 F. Supp. 109, 114 (D. Colo. 1979).

[9] *Id.*

[10] W.H. Lyman Constr. Co. v. Village of Gurnee, 403 N.E.2d 1325 (Ill. App. Ct. 1980).

[11] 415 N.E.2d 855 (Mass. App. Ct. 1981).

[12] *Id.* at 857.

[13] *Id.*

[14] Clark Bros. Contractors v. State, 710 P.2d 41 (Mont. 1985).

which the court determined not to be enough time to perform its own quantity estimates.[15]

§ 4.5 —Representations that Are Not Specific

Whether an owner has effectively disclaimed responsibility for its representations of subsurface conditions will depend not only on the strength of the disclaimer, but also on the nature of the representations made. When a contract contains an erroneous *positive* statement of conditions to be found, a contractor who relied on the statement may recover for breach of warranty.[16] As one court noted:

> [A] general disclaimer is ineffective to qualify a specific misleading representation; the general duty such a disclaimer imposes is superceded by a specific, positive misstatement, and the bidder is not required to search further for the facts.[17]

However, when specifications provided by an owner offer a contractor the results of investigations or tests, but do not purport to describe actual conditions or fairly disclaim any responsibility for such descriptions, a contractor may not assert a breach-of-warranty claim when the conclusion the contractor draws from the results turns out to be inaccurate.[18] The basis for the distinction lies in the fact that a contractor is not justified in relying on such information without independent investigation.

Positive representations of conditions to be encountered would include a contract's depictions of elevations. These would not simply be test results that might "have alerted a contractor for a need for further inquiry."[19] Another specific positive representation would be an owner's statement as to the amount of "unsuitable material" a contractor will have to excavate. When an owner made such a statement, instead of offering the contractor the testing information from which such a conclusion could have been drawn, the contractor was able to recover the excess costs of the more extensive excavation actually required, even though the contract contained several exculpatory clauses.[20] Likewise, a representation of the size and dispersion of boulders that a contractor could expect to

---

[15] *Id.*

[16] E.H. Morrill Co. v. State, 65 Cal. 2d 787, 423 P.2d 551, 56 Cal. Rptr. 479 (1967); Golomore Assocs. v. New Jersey State Highway Auth., 173 N.J. Super. 55, 413 A.2d 361 (App. Div. 1980).

[17] Thorn Constr. v. Utah Dep't of Transp., 598 P.2d 365, 367–68 (Utah 1979).

[18] Wunderlich v. State, 65 Cal. 2d 777, 423 P.2d 545, 56 Cal. Rptr. 473 (1967).

[19] Golomore Assocs. v. New Jersey State Highway Auth., 173 N.J. Super. 55, 413 A.2d 361, 362 (App. Div. 1980).

[20] Alpert v. Commonwealth, 258 N.E.2d 755 (Mass. 1970).

encounter constituted a positive assertion of fact of conditions that were not effectively disclaimed to avoid a breach-of-warranty claim based on the different conditions actually encountered.[21]

When, however, instead of making positive statements of actual conditions, an owner sets forth the results of tests or investigations, the owner is "simply under an obligation to make no false statement."[22] If the owner accurately sets forth the results, the assumptions drawn by a contractor are not warranted or guaranteed.[23] The statements made are "suggestive" or "indications," and expenses caused by unforeseen conditions will be borne by the contractor, especially if the owner had disclaimed responsibility for the conditions.[24]

§ 4.6 Misrepresentation

A contractor should be aware that a breach-of-warranty claim is not the only method by which a contractor may seek recovery against an owner when subsurface conditions are not as indicated in a contract. Another theory that may be advanced is one based on misrepresentation. An owner's misrepresentation of subsurface conditions may render an otherwise valid disclaimer ineffective. Even when a court may find that the owner's representations did not constitute an implied warranty, it may nevertheless hold an owner liable to a contractor for misrepresenting facts about the soils conditions.[25]

Claims based on misrepresentation are very similar to claims based on breach of warranty, and the cases seem to have been generated from the same theory, that is, a contractor should be able to accept as true an owner's positive statements in a contract, and the owner should be responsible for the expenses caused by their inaccuracy.[26]

§ 4.7 —Innocent Misrepresentation

Contrary to the requirements that must be established in a classic misrepresentation claim, in most cases based on misrepresentation of subsurface conditions, the courts do not look to whether the owner tried to deceive the contractor into relying on data the owner knew to be false. The Supreme Court has long held

---

[21] E.H. Morrill Co. v. State, 65 Cal. 2d 787, 423 P.2d 551, 56 Cal. Rptr. 479 (1967).

[22] Wunderlich v. State, 65 Cal. 2d 777, 423 P.2d 545, 549, (56 Cal. Rptr. 473 citing Chris Nelsen & Son v. City of Monroe, 337 Mich. 438, 60 N.W.2d 182 (1953)).

[23] 423 P.2d at 549.

[24] *Id.*

[25] Eastern Tunneling Corp. v. Southgate Sanitation Dist., 487 F. Supp. 109 (D. Colo. 1979); D. Federico Co. v. Commonwealth, 415 N.E.2d 855 (Mass. App. Ct. 1981).

[26] Hollerbach v. United States, 233 U.S. 165 (1914).

that whether there was a "sinister purpose" is not material in a case based on a misrepresentation of soils conditions.[27] Instead, in assessing the reasonableness of the reliance by a contractor, the test applied by most courts facing a claim of misrepresentation of subsurface conditions is (1) whether the contractor could have discovered the correct facts concerning the soils conditions through reasonable investigation before the contract, and (2) whether the data actually provided by the owner was inaccurate.[28] The presence of even specific disclaimers will not necessarily preclude a misrepresentation claim.[29]

In many jurisdictions, an owner's disclaimers will not preclude a contractor from justifiably relying on statements in a contract and presenting a claim for misrepresentation. The inquiry will be made as to whether the contractor's reliance was reasonable under all the circumstances.[30] For example, if a site inspection was required but would not have given a contractor more accurate information than that contained in the bid documents, a disclaimer will not prevent a claim based on their errors.[31] A disclaimer will not prevent recovery when the contractor made a reasonable site inspection,[32] and the contractor will not be required to have made independent subsurface investigations or soils analyses in order to recover.[33]

In government contracts, the failure of the contractor to conduct independent subsurface exploration will not prevent recovery for misrepresented soils conditions when the time between the advertisement for bids and submission of bids was insufficient for the contractor to make such a study,[34] or when the physical conditions at the time bids were prepared made such an investigation impossible.[35] These factors will determine whether the owner actually expected the contractor to make an independent investigation or whether reliance on the contract indications was anticipated.[36]

In some jurisdictions, however, the presence of disclaimers in the contract, by themselves, makes it impossible for a contractor to argue that the owner should

---

[27] Christie v. United States, 237 U.S. 234 (1915).

[28] Eastern Tunneling Corp. v. Southgate Sanitation Dist., 487 F. Supp. 109, 115 (D. Colo. 1979).

[29] Fehlhaber Corp. v. United States, 138 Ct. Cl. 571, 151 F. Supp. 817 (Ct. Cl.), *cert. denied,* 355 U.S. 877, 78 S. Ct. 141 (1957).

[30] J.F. Shea Co. v. United States, 4 Cl. Ct. 46, 53 (1983).

[31] Continental Drilling, ENG BCA No. 3455, 75-2 B.C.A. (CCH) ¶ 11,541 (1975).

[32] Baltimore Contractors, Inc. v. United States, 12 Cl. Ct. 328 (1987).

[33] County Asphalt v. State, 40 A.D.2d 26, 337 N.Y.S.2d 415 (1972).

[34] Christie v. United States, 237 U.S. 234 (1915); Fehlhaber Corp. v. United States, 138 Ct. Cl. 571, 151 F. Supp. 817 (Ct. Cl.), *cert. denied,* 355 U.S. 877, 78 S. Ct. 141 (1957).

[35] Al Johnson Constr. Co. v. Missouri Pac. R.R. Co., 426 F. Supp. 639 (E.D. Ark. 1976), *aff'd,* 553 F.2d 103 (8th Cir. 1977).

[36] *Id.*

be liable for misrepresenting subsurface conditions.[37] For example, an owner was not held responsible for misrepresentations of site conditions when the contract stated that certain data were not to be relied upon and were not to be considered part of the contract. In light of these provisions, the court held that the contractor did not justifiably rely on the data.[38] The question of whether the contractor had enough time to do an independent investigation was not even considered.[39]

Moreover, whether a contractor's reliance on the owner's representations is reasonable under the circumstances will depend not only on the owner's conduct but also on the contractor's. A contractor was unable to prove that its reliance on the contract documents was reasonable when it had failed to review information that the bid documents identified as available and when its own test results and its prior experience in that location should have put it on notice that the contract indications of soils conditions were erroneous.[40]

§ 4.8 —Intentional Misrepresentation

As examined so far, there are many factors to be considered in analyzing the effect of a disclaimer on the risk of unforeseen site conditions. However, when an owner has intentionally misrepresented the conditions to the contractor, little analysis is required to determine that the owner's disclaimer of responsibility will not be held to be valid. When the contractor can establish "intentional" or "conscious" misrepresentation or fraud (a difficult burden to be sure), the owner's disclaimer will not preclude recovery.[41] If an owner knowingly makes misrepresentations on which it intends a contractor to rely, it cannot also insulate itself from liability based on disclaimers.[42]

An owner cannot simply disclaim liability for conditions that it knows to be inconsistent with its contract documents. A flagrant example of such an attempt is found in a case in which two different contractors had abandoned work on a sewer construction project when the rock formation indicated in the contract had not been encountered. Before letting a contract to complete the work, the owner simply appended a disclaimer to the contract drawings. When the third

[37] J.E. Brenneman Co. v. Commonwealth, 424 A.2d 592 (Pa. Commw. Ct. 1981).

[38] *Id.; see also* McDevitt & St. Co. v. Marriott Corp., 713 F. Supp. 906 (E.D. Va. 1989); Jahnke Serv., Inc. v. Department of Transp., 322 S.E.2d 505 (Ga. Ct. App. 1984).

[39] *Id.; see also* S&M Constructors v. City of Columbus, 434 N.E.2d 1349 (Ohio 1982).

[40] Umpqua River Navigation Co. v. Crescent City Harbor Dist., 618 F.2d 588 (9th Cir. 1980).

[41] Husman, Inc. v. Triton Coal Co., 809 P.2d 796 (Wyo.), *reh'g denied,* 191 Wyo. LEXIS (Wyo. 1991).

[42] Passaic Valley Sewerage Comm'rs v. Holbrook Cabot & Rollins Corp., 6 F.2d 721 (3d Cir.), *cert. denied,* 269 U.S. 582, 46 S. Ct. 107 (1925).

contractor made a claim based on the subsurface conditions, the disclaimer did not relieve the owner of its responsibility for "conscious misrepresentations."[43]

That an owner "knew or should have known" that its representations of subsurface conditions were untrue precluded enforcement of disclaimers in an Illinois case.[44] There, a sewer construction contract contained numerous specific disclaimers of responsibility for the presence of underground structures. The contractor was nevertheless able to recover for the additional costs incurred when it encountered an elaborate and extensive drainage tile collection system not shown in the contract. Although the plans purported to show "a very large percentage" of the underground utilities, in fact, only a small percentage of feet of drainage tile had been identified. Although the agreement not only required the contractor to assume all risks of subsurface conditions encountered, but also to restore underground utilities not shown on the plans, the contractor nevertheless recovered its additional expenses. The court found that the municipal project owner "knew or should have known" of the drainage system because it had passed an ordinance addressing the issue a few years before the project had been bid.[45]

A principle related to the owner's duty not to misrepresent site conditions is the duty of the owner to disclose information to the contractor. The issue of "superior knowledge" arises in government contract cases in which a contractor may argue that the government possessed special knowledge or data vital to the performance of the contract but which was unknown or unavailable to the bidder who was thereby misled.[46] Thus, the state of New York was unable to disclaim liability for misrepresentations of site conditions relied upon by a bidder because the actual conditions were found to be "peculiarly within the knowledge of the State":

> It is a well established doctrine that a party to a contract cannot by a misrepresentation of a material fact induce the other party to enter into it to his damage, and then protect himself from the legal effect of such misrepresentation by inserting a clause to the effect that: 'He is not to be held liable for the misrepresentation which induced the other party to enter into the contract. The effect of misrepresentation and fraud cannot be thus easily avoided. If it could be, the implied covenant of good faith and fair dealing existing in every contract would cease to exist.'[47]

---

[43] *Id.*

[44] Horton Indus., Inc. v. Village of Moweaqua, 492 N.E.2d 220 (Ill. App. Ct. 1986).

[45] *Id.* at 227.

[46] *See* Aleutian Constructors v. United States, 24 Cl. Ct. 372, 383 (1991); J.F. Shea Co. v. United States, 4 Cl. Ct. 46, 53 (1983); Helene Curtis Indus., Inc. v. United States, 160 Ct. Cl. 437, 442, 312 F.2d 774, 777 (1963).

[47] Young Fehlhaber Pile Co. v. State, 177 Misc. 204, 206, 30 N.Y.S.2d 192, 195 (Ct. Cl. 1941), *aff'd,* 265 A.D. 61, 37 N.Y.S.2d 928 (1942).

To best avoid liability for claims of intentional misrepresentation, an owner should disclose any relevant information it has. If a contractor can show that the owner concealed relevant information regarding subsurface conditions, the contractor's recovery will not be barred by disclaimers.

§ 4.9 —Negligent Misrepresentation

Along with breach of warranty and "innocent" and fraudulent misrepresentation, some courts will render an otherwise valid disclaimer ineffective if the contractor can prove that the owner negligently misrepresented the subsurface conditions in the bid or contract documents.[48] The definition and/or the elements of negligent misrepresentation vary from jurisdiction to jurisdiction.

Negligent misrepresentation has been defined as the positive assertion, in a manner not warranted by the information of the person making it, of that which is not true though he believes it to be true, to induce a party to enter into a contract.[49] A statement made for the guidance of others that is not warranted by the information of the person making it is classified as negligent misrepresentation.[50] Generally, the elements of negligent misrepresentation include:

1. A positive, false misrepresentation of material fact
2. Such misrepresentation supplied in the course of one's business for the guidance of others in their business
3. A failure to exercise reasonable care in obtaining or relating the information
4. Pecuniary loss resulting from justifiable reliance thereon.[51]

Unlike fraudulent misrepresentation, to prove negligent misrepresentation, the contractor is not required to show that the owner had made a positive averment of site conditions with the *intent* of deceiving the contractor.[52]

Notwithstanding the foregoing, the misrepresentation must be based on a positive assertion of subsurface conditions. If the subsurface information furnished

---

[48] *See, e.g.,* Murphy v. City of Springfield, 738 S.W.2d 521 (Mo. Ct. App. 1987); Western States Mechanical Contractors, Inc. v. Sandia Corp., 110 N.M. 676, 798 P.2d 1062 (1990); Davidson & Jones, Inc. v. County of New Hanover, 41 N.C. App. 661, 255 S.E.2d 580 (1979); Bourgois v. Montana-Dakota Utils. Co., 466 N.W.2d 813 (N.D. 1991); Husman, Inc. v. Triton Coal Co., 809 P.2d 796 (Wyo.), *reh'g denied,* 191 Wyo. LEXIS (Wyo. 1991).

[49] Bourgois v. Montana-Dakota Utils. Co., 466 N.W.2d 813 (N.D. 1991).

[50] *Id.* at 817.

[51] Husman, Inc. v. Triton Coal Co., 809 P.2d 796 (Wyo.), *reh'g denied,* 191 Wyo. LEXIS (Wyo. 1991); Murphy v. City of Springfield, 738 S.W.2d 521, 527 (Mo. Ct. App. 1987).

[52] In almost all jurisdictions, courts will not uphold an exculpatory clause if the contractor can prove that the owner *intentionally* misrepresented subsurface conditions in the bid documents.

by the owner amounts to nothing more than implicit suggestions of the subsurface conditions, such inferential conclusions will not be considered a false factual representation and will not overcome a valid exculpatory clause.[53]

Most courts today require that the contractor make its own reasonable independent investigation of the site notwithstanding the conditions described in the contract documents.[54] Although the government cannot make a contractor the insurer of all government mistakes,[55] it has been held that the contractor must absorb expenses that would have been avoided if the contractor had been conscientious in its investigation.[56] A contractor may not rely on contract indications of the subsurface conditions when a relatively simple inquiry might have revealed contrary conditions.[57]

In *Husman, Inc. v. Triton Coal Co.,*[58] after two of its employees inspected a mine site, Husman, the contractor, entered into an agreement with Triton, the mining company, to remove certain amounts of overburden and topsoil at the site. Before Husman's inspection of the site, Triton made a representation to Husman that the overburden was "an unconsolidated material and . . . it was probably a sandy clay type material."[59] After Husman began working, it discovered that the overburden was saturated with moisture and was more difficult to remove than it had anticipated.

After the contract was terminated, Husman commenced an action against Triton seeking damages for the additional costs of subsoil removal based on Triton's negligent misrepresentation of the subsoil conditions. Following Oklahoma law, the court held that the exculpatory clause did not bar Husman from asserting its negligent misrepresentation claim because a contract clause that requires a contractor to rely upon its own inspection does not control when there is a

---

[53] P.T.& L. Constr. Co. v. State, 108 N.J. 539, 531 A.2d 1330 (1987).

[54] *See* D. Federico Co. v. New Bedford Redev. Auth., 723 F.2d 122, 125 (1st Cir. 1983); Umpqua River Navigation Co. v. Crescent City Harbor Dist., 618 F.2d 588 (9th Cir. 1980); Western States Mechanical Contractors, Inc. v. Sandia Corp., 110 N.M. 676, 798 P.2d 1062, 1065 (1990); Husman, Inc. v. Triton Coal Co., 809 P.2d 796, 800 (Wyo.), *reh'g denied,* 191 Wyo. LEXIS (Wyo. 1991).

This requirement arises from the fact that most, if not all, of today's excavation contracts impose a duty on the bidder-contractor to familiarize itself with the site conditions before bidding. Aside from the standard disclaimers disclaiming the accuracy of subsurface information, contracts containing excavation work usually also contain clauses that impose a duty on the bidder-contractor to perform its own pre-bid site investigation and/or take its own test borings. If no such affirmative duty is imposed, the contract will usually contain a statement whereby the contractor acknowledges that it has investigated the site and/or reviewed any and all materials that would indicate the subsurface conditions.

[55] Michigan-Wis. Pipeline Co. v. Williams-McWilliams Co., 551 F.2d 945, 953 (5th Cir. 1977).

[56] D. Federico Co. v. New Bedford Redev. Auth., 723 F.2d 122, 125 (1st Cir. 1983).

[57] Umpqua River Navigation Co. v. Crescent City Harbor Dist., 618 F.2d 588, 594 (9th Cir. 1980).

[58] 809 P.2d 796 (Wyo.), *reh'g denied,* 191 Wyo. LEXIS (Wyo. 1991).

[59] *Id.* at 798.

finding of misrepresentation as to existing conditions. It is not unreasonable for a contractor to rely upon the owner's representations and, therefore, limit its investigation.[60]

In its decision, the court did not uphold Husman's claim. The court only concluded that it might have been reasonable for Husman to rely on Triton's representation. To determine if Husman should recover, the trier of fact would have to decide if Husman could have discovered the true subsurface conditions upon its own *reasonable* site inspection. The court remanded this question back to the trial court for a determination.

Whether the contractor can discover the true facts through a reasonable site investigation depends on the facts of each case.[61] In deciding whether the contractor has made a reasonable inspection of the site, many courts will not require the contractor to spend large sums of money by taking the same test borings taken by the owner.[62] In *Jacksonville Port Authority v. Parkhill-Goodloe Co.,*[63] the court recognized that it would be inefficient and unfair to require each bidder to run tremendously expensive test borings without knowing whether or not it would get the contract. On the other hand, the county (owner) should know what soil it is dealing with in a project under consideration. The information, if gathered by the county (owner), would be useful to the county no matter who gets the job. The county (owner) should be the logical one to make the tests or have them made and not the bidders who may or may not get the job. The bidders should be able to rely on these tests in all justice and fair play.[64]

Courts will also take into account the reasonableness of the contractor's interpretation and reliance of its own pre-bid subsurface investigations. For example, in *Umpqua River Navigation Co. v. Cresent City Harbor District,*[65] the court held that although the owner provided the contractor with inaccurate test borings, the contractor should not have relied on them in light of its own test borings that were inconsistent with those provided by the owner. This inconsistency should have put the contractor on notice that the contractual indications were suspect.

It is evident that a prudent contractor should perform an independent inspection of the site when the bid documents impose a duty to inspect and/or

---

[60] *Id.* at 800.

[61] Western States Mechanical Contractors, Inc. v. Sandia Corp., 110 N.M. 676, 798 P.2d 1062, 1065 (1990).

[62] Jacksonville Port Auth. v. Parkhill-Goodloe Co., 362 So. 2d 1009 (Fla. 1978); *see also* Robert E. McKee, Inc. v. City of Atlanta, 414 F. Supp. 957 (N.D. Ga. 1976).

[63] 362 So. 2d 1009 (Fla. 1978).

[64] *See also* Condon-Cunningham, Inc. v. Day, 22 Ohio Misc. 71, 258 N.E.2d 264, 51 Ohio Op. 2d 144 (1969).

[65] 618 F.2d 588, 594 (9th Cir. 1980) (citing Foster Constr. C.A. v. United States, 435 F.2d 873, 193 Ct. Cl. 587 (1970)).

contain any type of disclaimer regarding misrepresentation of subsurface conditions. If such an inspection indicates subsurface conditions inconsistent with those described in the bid materials, the contractor should seek further data from the owner and/or perform a more extensive subsurface inspection before submitting its bid. Failure to do so may preclude a contractor from recovering damages on a claim based on a negligent misrepresentation of subsurface conditions.

The omission of material subsurface information may also give rise to a negligent misrepresentation of subsurface conditions. For example, in *Bourgois v. Montana-Dakota Utils. Co.,*[66] the North Dakota Supreme Court held a general exculpatory clause invalid which essentially stated that the contractor acknowledged that it would perform demolition and cleanup work as delineated in the contract, and that the contract price was the maximum compensation to be paid to the contractor notwithstanding any unforeseen difficulties.

During demolition, Bourgois, the contractor, uncovered large blocks of buried concrete. Bourgois brought an action against the owner claiming, among other things, damages for extra work for removing the buried concrete based on the owner's negligent failure to disclose the buried concrete. The owner sought dismissal of Bourgois's action on the ground that the exculpatory clause prevented Bourgois from receiving compensation over and above the contract price. In opposing the owner's motion for the dismissal of Bourgois's action, Bourgois alleged that neither the plans nor the specifications furnished by the owner disclosed the buried concrete.

During the litigation, Bourgois found an architect who had a copy of plans showing the buried concrete. These plans had been delivered by the owner to the architect approximately 10 years before Bourgois's work at the site. The court ruled that even though Bourgois could not prove that the owner knew of the existence of the buried concrete at the time Bourgois's contract was negotiated, the owner had negligently failed to disclose the presence of the buried concrete. As did the courts in *Jacksonville Port Authority v. Parkhill-Goodloe Co.* [67] and *Condon-Cunningham, Inc. v. Day* ,[68] the court decided that the owner was in a better position to discover and disclose the buried concrete.

Even though the United States Supreme Court has held that exculpatory clauses do not necessarily preclude liability,[69] not all disclaimers will be held invalid in the face of a negligent misrepresentation of subsurface conditions. There are cases in which the court held that although a general exculpatory clause would not be upheld due to a negligent misrepresentation of subsurface conditions, a disclaimer *specifically* disclaiming responsibility for the contested information would nevertheless absolve the owner for the inaccuracy of

---

[66] 466 N.W.2d 813 (N.D. 1991).

[67] 362 So. 2d 1009 (Fla. 1978).

[68] 22 Ohio Misc. 71, 258 N.E.2d 264, 51 Ohio Op. 2d 144 (1969).

[69] Hollerbach v. United States, 233 U.S. 165, 172 (1914).

positive and material representations upon which the contractor had a right to rely.[70]

For example, in *Western States Mechanical Contractors, Inc. v. Sandia Corp.,*[71] the contractor claimed that it suffered damages because the owner negligently misrepresented the amount of rock to be removed. In the contract, the owner disclaimed the information contained in a soils report prepared for it by a soils engineer. The court held that although the soils report may have been disclaimed, the amount of rock as represented in the plans and specifications was not disclaimed. The court did not agree with the owner's argument that the rock estimate was an interpretation of the soils report.

There are also jurisdictions that uphold valid disclaimers and preclude claims based on negligent misrepresentation of subsurface conditions altogether. There are jurisdictions that have statutory prohibitions against claims for negligent misrepresentations,[72] while others base their decision in accordance with common law rules of contract interpretation.[73] In many cases, some jurisdictions simply do not recognize such a claim.[74]

For example, in *Berkel & Co. Contractors, Inc. v. Providence Hospital,*[75] the Alabama Supreme Court denied a contractor's claim that it relied on the owner's negligently misrepresented subsurface soil conditions in preparing its bid. The court based its decision on a valid exculpatory clause in the bid documents that disclaimed liability for the adequacy or content of the owner's soils report. The facts in *Berkel* are instructive. Berkel was hired by the general contractor to install piles for the foundation of an addition to Providence Hospital in Mobile, Alabama. When the owner issued an invitation for bids, it made available to the bidders the plans, specifications, and addendum prepared by its architect and a soils report prepared for the owner by an engineering firm. The soils report was based on eight soil test borings, only one of which was taken in the actual excavation site. The subcontract incorporated the bidders' information package by reference.

When Berkel installed two test piles, both failed a load test. Nevertheless, Berkel received instructions from the architect through the general contractor to proceed with the installation of the production piles. At the same time, Berkel was directed to install another test pile, which failed. By the time Berkel

---

[70] P.T.& L. Constr. Co. v. State, 108 N.J. 539, 531 A.2d 1330 (1987); Western States Mechanical Contractors, Inc. v. Sandia Corp., 110 N.M. 676, 798 P.2d 1062, 1065 (1990).

[71] 110 N.M. 676, 798 P.2d 1062 (1990).

[72] *See, e.g.,* Wiechmann Eng'rs v. State, 31 Cal. App. 3d 741, 107 Cal. Rptr. 529 (1973). The relevant California statute concerns only negligent misrepresentations by public employees. Thus, it is not applicable to negligent misrepresentations by private entities.

[73] *See, e.g.,* Berkel & Co. Contractors, Inc. v. Providence Hosp., 454 So. 2d 496 (Ala. 1984).

[74] *See, e.g.,* Industrial Dredging & Eng'g Corp. v. Southern Ind. Gas & Elec. Co., 840 F.2d 523 (7th Cir. 1988) (no such claim recognized under Indiana law).

[75] 454 So. 2d 496 (Ala. 1984).

had installed 87 of the production piles, it was directed to cease any more installation.

The contract contained a disclaimer essentially stating that the soils reports had been prepared by an engineer for the owner and that no representation or warranty was made by the architect/engineer or owner regarding the adequacy or content of the reports. Moreover, the disclaimer required bidders to obtain, at their own expense, information necessary to determine existing site and subsoil conditions to perform the contract completely as specified in a proper and safe manner.

In support of its claim, Berkel contended that the disclaimer should be ineffective on policy grounds. Berkel cited *Condon-Cunningham, Inc. v. Day* for the proposition that the owner will be held liable when the contractor reasonably relies on the furnished information.[76] Berkel presented affidavits attesting to the industry practice of relying on owner-furnished soils information in preparing bids because the owner is in the best position to provide the information. It argued that allowing the owner to disclaim responsibility could have the economic effect of increasing the cost of bids to reflect the risk shift of a disclaimer, incurring the prospect of repetitious tests, and reducing the number of bidders.[77]

The court held that despite the economic policy objections, Alabama law embraces the freedom of contract. The marketplace, not the judiciary, is the proper forum to determine the costs and inefficiencies of the construction. The exculpatory clause clearly and unequivocally disclaimed responsibility for reliance on the soils report. The thoroughness of the soils report disclaimer satisfied the elements of a legally effective disclaimer. The court took into account the fact that the disclaimer (1) provided that the soils report was prepared solely for the owner and was not meant to be relied on by third parties, and (2) included the necessary step of requiring the bidder to obtain information at its expense to satisfy itself.[78] Hence, the owner had successfully disclaimed liability for the negligence of its agent (the engineer).

Although exculpatory clauses disclaiming any liability for misrepresentations are generally enforceable, there are jurisdictions that will render an otherwise valid disclaimer ineffective if the owner negligently misrepresents the subsurface conditions. Negligent misrepresentation may be proved by showing that the owner made a positive and material misrepresentation as to subsurface conditions when the facts known to it did not warrant the making of the representation. Notwithstanding the foregoing, it is clear that when the contract imposes a duty to investigate, most jurisdiction will require the contractor to show that it

---

76 22 Ohio Misc. 71, 258 N.E.2d 264, 51 Ohio Op. 2d 144 (1969). The court refused to recognize the policies that the court in *Condon-Cunningham, Inc.* used as the reasoning behind holding the disclaimer ineffective.

77 Berkel & Co. Contractors, Inc. v. Providence Hosp., 454 So. 2d 496, 504 (Ala. 1984).

78 *Id.*

made a reasonable inspection of the site to recover damages on its claim. Additionally, an alleged positive averment of subsurface conditions must be more than suggestive. It must be an actual description of subsurface conditions that the contractor relied on in preparing its bid.

No matter what type of exculpatory clause is included in the bid or contract documents, each claim for negligent misrepresentation must be reviewed on a case-by-case basis. Because jurisdictions are split as to whether, and/or when, a negligent misrepresentation of subsurface conditions will render an exculpatory clause ineffective, each claimant must research its jurisdiction's policy regarding negligent misrepresentation of subsurface conditions as a basis for rendering a valid disclaimer ineffective.

CHAPTER 5

EFFECT OF SITE INVESTIGATION CLAUSES ON CLAIMS

John E. Schulz

John E. Schulz is of counsel to the environmental litigation and counseling group at Bronson, Bronson and McKinnon, San Francisco, California. He has an unusual wealth of environmental legal experience, including compliance counseling, advising on transfers of contaminated real estate, plaintiff and defense litigation, government affairs work, contract negotiations, writing and teaching, and legal seminar presentations. A Graduate of Princeton University and Yale Law School (Law Journal, Coif), Mr. Schulz previously did Superfund litigation for the U.S. Department of Justice; carried out contracting, risk evaluation, government affairs, and seminar work for Bechtel Environmental, Inc.; and served as Editor in Chief of the *Environmental Law Reporter.* He is a member of the American Bar Association Natural Resources Section and of the Executive Committee of the State Bar of California Environmental Law Section.

---

§ 5.1 Introduction

As discussed in earlier chapters, differing site conditions clauses provide for making cost adjustments between contractors and owners in response to increases or decreases in anticipated performance costs. Their function is to shift some of the performance risks of latent or unknown subsurface conditions from contractor to owner.[1] Site investigation clauses, which require contractors to perform site investigations before bidding, shift some of that same risk back from owner to contractor.

In theory, unexpected differences in latent or subsurface conditions must in some cases *decrease* rather than increase the cost of performance. In such a situation, however, a contractor is not likely to make a claim, and the owner may not be aware of the difference despite the "notice" requirement in typical differing site conditions clauses.[2] As a result, there appear to be no published cases dealing with "reverse" differing site conditions claims.[3] Accordingly, the case law deals with contractors making differing site conditions claims to owners, or subcontractors to general contractors, rather than the reverse.

This chapter addresses the impact of site investigation clauses on the validity of claims under differing site conditions clauses. Differing site conditions claims are of two kinds: Type I claims refer to conditions in the subsurface area that are materially different from those stated in the contract; Type II claims are those that relate to unknown conditions of an unusual nature, not ordinarily encountered in the kind of work provided for in the contract. In Type II cases, the contract typically contains no language describing the conditions at issue.

Many cases construing and applying differing site conditions clauses and site investigation clauses and analyzing their interaction are based on interpretations of such clauses provided for in the Federal Acquisition Regulation (FAR).[4] Most

---

[1] The common law rule imposes the risk of encountering and coping with unexpected conditions on the contractor. *See* United States v. Spearin, 248 U.S. 132 (1918); Dravo Corp. v. Litton Sys., 379 F. Supp. 37 (D. Miss. 1974).

[2] *See* FAR 52.236-2, 48 C.F.R. § 52.236-2 (1984).

[3] LEXIS searches in the MEGA file of the GENFED and STATES Libraries and the CTSBCA file of the PUBCON Library on the topic "(CHANGED OR CONCEALED OR UNKNOWN OR DIFFERING) PRE/5 CONDITION! W/30 DECREASE! PRE/5 (COST! OR EXPENSE! OR TIME OR CONTRACT PRE/2 SUM)" consistently turned up no cases.

[4] *See* FAR 52.236-2, 48 C.F.R. § 52.236-2 (1984) (FAR differing site conditions clause) and FAR 52.236-3, 48 C.F.R. § 52.236-3 (1984) (FAR site investigation clause), respectively. The FAR differing site conditions clause provides, in relevant part:

> (a) The Contractor shall promptly, and before the conditions are disturbed, give a written notice to the Contracting Officer of:
>
> > (1) subsurface or latent physical conditions at the site which differ materially from those indicated in this contract, or
> >
> > (2) unknown physical conditions at the site, of an unusual nature, which differ materially from those ordinarily encountered and generally recognized as inhering in work of the character provided for in the contract.
>
> (b) The Contracting Officer shall investigate the site conditions promptly after receiving the notice.

The FAR site investigation clause provides:

> (a) The Contractor acknowledges that it has taken steps reasonably necessary to ascertain the nature and location of the work, including but not limited to (1) conditions bearing upon transportation, disposal, handling, and storage of materials; (2) the

case law in this area involves federal public works contracts, and the most prolific source is various boards of contract appeals (BCA) decisions.

In addition, the 1987 edition of the American Institute of Architects (AIA) Document A201 essentially adopts the provisions of the FAR differing site conditions and site investigation clauses.[5] Accordingly, the body of federal precedent on these clauses is now available to guide state courts in interpreting and applying the current AIA clauses in private contract cases.

Typically, contracts that contain a differing site conditions clause also contain a site investigation clause. This is required by the FAR in federal public works contracts.[6] FAR 36.503 requires insertion of the FAR site investigation clause in all public contracts that require the use of the FAR differing site conditions clause. The site investigation clause is the contractor's acknowledgment that it has conducted a site investigation and is satisfied with the conditions.

To understand the typical impact of the FAR and AIA site investigation clauses, it is helpful to set forth the requirements for proving both a Type I and Type II differing site conditions claim. In the words of the FAR differing site conditions clause, a Type I claim is one for:

> availability of labor, water, electric power, and roads; (3) uncertainties of weather, river stages, tides, or similar physical conditions at the site; (4) the conformation and the conditions of the ground; and (5) the character of equipment and facilities needed preliminary to and during work performance. The Contractor also acknowledges that it has satisfied itself as for the character, quality, and quantity of the surface and subsurface materials or obstacles to be encountered insofar as this information is reasonably ascertainable from an inspection of the site; including all exploratory work done by the government, as well as from the drawings and specifications made a part of this contract. Any failure of the Contractor to take the actions described and acknowledged in this paragraph will not relieve the Contractor from responding for estimating properly the difficulty and cost of successfully performing the work, or for proceeding to successfully perform the work without additional expense to the government.

See also 23 C.F.R. § 635.131(a) (1991), which sets forth the differing site conditions clause of the Federal Highway Commission for highway construction and maintenance projects.

[5] *See* AIA A201-1987, General Conditions, ¶¶ 4.3.6 (differing site conditions clause):

> Should concealed conditions encountered in the performance of the work below the surface of the ground or should concealed or unknown conditions in an existing structure be at variance with the conditions indicated by the Contract Documents, or should unknown physical conditions below the surface of the ground or should concealed or unknown conditions in an existing structure of an unusual nature, differing materially from those ordinarily encountered and generally recognized and inherent in work of the character provided for in this Contract, be encountered, the Contract Sum shall be equitably adjusted by Change Order upon claim by either party made within 20 days after the first observance of the conditions.

and 1.2.2 (site investigation clause).

> By executing this Contract, the Contractor represents that he has visited the site, familiarized himself with the local conditions which the work is to be performed, and correlated his observations with the requirements of the Contractor documents.

[6] FAR 36.503, 48 C.F.R. § 36.503 (1984).

(1) subsurface or latent physical conditions at the site which differ materially from those indicated in this contract.

A Type II claim is for:

(2) unknown physical conditions at the site, of an unusual nature, differing materially from those ordinarily encountered and generally recognized as inhering in a work of the character provided for in this contract.

§ 5.2 Type I Claim Requirements

Recently, the Department of Veterans Affairs Board of Contract Appeals set forth a six-element test to use when reviewing Type I claims. This test fleshes out the language of the FAR provision. The contractor must establish that:

1. Contract documents affirmatively indicate the subsurface or latent conditions forming the basis of the claim
2. The contractor acted as a reasonable and prudent contractor in interpreting the contract documents
3. The contractor relied on the indications of the subsurface or latent conditions in the contract
4. The subsurface or latent conditions encountered at the site differed materially from the contract indications
5. The actual conditions encountered were not reasonably foreseeable
6. The costs claimed were solely attributable to materially different subsurface conditions.[7]

§ 5.3 Type II Claim Requirements

A Type II claim requires the contractor to demonstrate that it has encountered an unknown or unusual condition that is materially different from that ordinarily encountered.[8] In the Type II situation, there are no representations concerning physical conditions in the language of the contract whatsoever; the contract is silent as to subsurface or latent conditions. This silence need not be as to *all* conditions at the site, but there must be no written reference at all to the conditions that form the basis for the Type II claim.

---

[7] Simpson Constr. Co., VABCA No. 3,176, 91-1 B.C.A. (CCH) ¶ 23,630, at 118,390 (1990).

[8] *See, e.g.,* Imbus Roofing Co., GSBCA No. 10,430, 91-2 B.C.A. (CCH) ¶ 23,820, at 119,348 (1991).

In short, to prevail in a Type II case, a contractor must establish the following elements:

1. Conditions that were recognized and usual conditions at the site
2. The unknown physical conditions that the contractor actually encountered
3. The physical conditions actually encountered differed materially from the known and usual
4. The actual conditions caused an increase in the cost of performance.[9]

As discussed in §§ **5.4** and **5.5**, the burden of proof in Type II cases is heavier than in Type I cases.

§ 5.4 Site Investigation in Type I Cases

In Type I cases, site investigation clauses appear frequently to have less effect on differing site conditions clauses in governmental cases than in private contracts. It is, however, difficult to generalize. Specific cases often turn on factual details, as indicated by cases discussed later in this section.

The presence of a site investigation clause functions to create a duty on the contractor's part to investigate the site.[10] Failure to conduct any site investigation whatsoever may be fatal to a differing site conditions claim; however, it has also been held that failure of a contractor to conduct a site investigation when required to do so by the contract did not preclude a recovery on a claim as long as the contractor could demonstrate that a site investigation would not have revealed the true condition of the site in any event.[11]

Moreover, if recovery is denied by reason of the site investigation clause, a contractor may still have a common law claim against the owner's independent soils engineer or environmental consultant based on professional negligence.[12]

If the contractor does carry out a site investigation but fails to discover the latent or subsurface conditions, the contractor will have to demonstrate thatthe investigation conducted was reasonable.[13] This does not require the contractor

---

[9] *See* Charles T. Parker Constr. Co. v. United States, 433 F.2d 771, 778 (Ct. Cl. 1970); Sandwich Islands Constr. Inc., ASBCA No. 35,244, 88-3 B.C.A. (CCH) ¶ 21,143, at 106,739 (1983).

[10] *See, e.g.,* Mojave Enter. v. United States, 3 Cl. Ct. 353 (1983).

[11] P.J. Maffei Bldg. Wrecking Corp., GSBCA No. 5,031, 80-2 B.C.A. (CCH) ¶ 14,647 (1980).

[12] *Cf.* M. Miller Co. v. Dames & Moore, 198 Cal. App. 2d 305, 18 Cal. Rptr. 13 (1961) (contractor defeated by owner's disclaimer not site investigation, clause allowed to sue owner's consultant).

[13] Simpson Constr. Co., VABCA No. 3,176, 91-1 B.C.A. (CCH) ¶ 23,630, at 118,390 (1990); Weeks Dredging & Contracting, Inc. v. United States, 13 Cl. Ct. 193 (1987), *aff'd,* 861 F.2d 728 (Fed. Cir. 1988).

to be a trained geologist or other expert, nor does it require that it hire such experts or conduct a full-blown technical investigation.[14]

The contractor will, however, have to demonstrate that its site investigation was reasonable and that such a reasonable investigation would not have turned up the conditions encountered.[15] The site investigation clause has been interpreted by the BCA to constitute a waiver only for conditions that could have been discovered by a reasonable site investigation.[16]

The Court of Claims has stated the rule of reasonable investigation as follows:

> In the cases arising under the modern changed conditions clause caution continues to be observed that the duty to make an inspection of the site does not negate the changed conditions clause by putting the contractor at peril to discover hidden subsurface conditions or those beyond the limits of inspection appropriate to the time available.
>
> The contractor is unable to rely on contract indications of the subsurface only where relatively simple inquiries might have revealed contrary conditions. When, for instance, logs in the contract show no water, as in the instant case, he need not seek out experts who may tell him of the possibility of subsurface water.[17]

As noted, the decisions in this area frequently turn on minor variances in fact situations. As a general matter, however, in the governmental work setting, both federal and state, differing site conditions clauses tend to be given effect as against site investigation clauses more frequently than in private contracting work. As articulated by the courts, the private party contracting setting is assumed to be more arm's length, with neither party having a fiduciary duty to the other. By contrast, in the government works area, it is considered unseemly for the government to take advantage of contractors by using inconsistent clauses[18] or withholding superior information.

---

[14] *See* Morrison-Knudsen v. United States, 178 Ct. Cl. 712, 345 F.2d 535 (1965); James Julian, Inc. v. President & Comm'r, 346 F.2d 205 (4th Cir. 1965).

[15] *See, e.g.,* Highland Constr. Co. v. Stevenson, 636 P.2d 1034 (Utah 1981) (Utah excavation subcontractor's claim failed when site since inspection would have revealed water-saturated soils and unstable road base).

[16] C&L Constr. Co., ASBCA No. 22,993, 81-1 B.C.A. (CCH) ¶ 14,943, *aff'd on reconsideration,* 81-2 B.C.A. (CCH) ¶ 15,373 (1981).

[17] Foster Constr. C.A. v. United States, 435 F.2d 873, 888 (Ct. Cl. 1970).

[18] *See, e.g.,* Frederick Snare Corp. v. Maine-New Hampshire Interstate Bridge Auth., 41 F. Supp. 638 (D.N.H. 1941), in which the court held that the differing site conditions clause overrode a conflicting clause in the contract requiring each bidder to inform itself fully of conditions relating to the construction at the time of opening of bids and stating that each bidder was presumed to have inspected the site. The court found that the contractor encountered subsurface conditions materially different from those shown in the plans or indicated in the specifications provided by the government. It also found that no accurate soundings could possibly have been done by the contractor within the short time allowed before bidding, and that this must have been known to the Bridge Authority's engineers who drafted the specifications. The court also

In *Longboat Key v. Carl E. Widell & Son,*[19] a state public works case, the contractor encountered excess water while constructing part of the central wastewater system for a town. The court held that the contractor could recover for the excess cost of dealing with this changed condition under the contract's differing site conditions clause. Despite the site investigation clause, the court observed that the differing site conditions clause was designed to provide for an adjustment between the public works contractor and the government because of increases or decreases in costs resulting from encountering conditions differing materially from those contemplated in the contract. The purpose of the clause was to assure that a bidder would not pad its bid to protect itself against loss resulting from unforeseen conditions, but would rely on the government unit's promise to grant it an equitable adjustment if the unforeseen occurred. The court found that when the conditions were, in fact, materially different from those contemplated by the contract, relief would not be denied because of the presence in the contract of one or more of various monetary exculpatory clauses, such as the site investigation clause requiring the contractor to examine the construction site.

In the private setting, where the AIA forms govern, *Crook v. Oklahoma Board of Public Affairs*[20] held that a contractor could not receive an equitable cost adjustment when no mention was made in its solicitation of an underground aquifer of which the owner knew. It was central to the Oklahoma court's ruling that the contractor had failed to attend a pre-bid conference at which the aquifer had been discussed with other potential bidders.

§ 5.5 Site Investigation in Type II Cases

In Type II cases, the conceptual impact of a site investigation clause is largely the same as in Type I cases. Failure to conduct a site investigation, as required in the federal public works contracts, waives the contractor's right to rely on the differing site conditions clause, but again only for those conditions that would be ascertainable by a reasonable site investigation.[21] In Type II cases,

---

refused to give effect to a clause in the specifications providing that the Bridge Authority did not guarantee data made available to the contractor from certain borings and would not be liable for subsurface conditions found during the work not corresponding with conditions as indicated by the data, and that should the contractor consider such data insufficient, it should make such investigation as it considered necessary, basing its bid upon its own opinion. The court observed that this language should not be held to apply in cases of substantial differences, as to do so would be a fraud upon the contractor.

[19] 362 So. 2d 719 (Fla. Dist. Ct. App. 1978).

[20] 736 P.2d 140 (Okla. 1987).

[21] C&L Constr. Co., ASBCA No. 22,993, 81-1 B.C.A. (CCH) ¶ 14,943, *aff'd on reconsideration,* 81-2 B.C.A. (CCH) ¶ 15,373 (1981). *See also* AAAA Enters., Inc., ASBCA No. 28,172, 86-1 B.C.A. (CCH) ¶ 18,628 (1985).

the contractor bears a heavier burden than in Type I cases of proving that a reasonable inspection would not have revealed the condition.[22]

Illustrating this doctrine are cases such as *Lyburn Construction Co.,*[23] in which the contractor failed to inspect. The court held that a reasonable inspection would have revealed unusually thick, dry paint that would have to be removed. However, in *Leiden Corp.,*[24] lack of evidence on the record as to whether the contractor conducted a site investigation was not held to bar its claim.

Occasionally, other reasons appear in the Type II cases to excuse contractors' failure to investigate. One is government action or inaction that was shown to be responsible for the failure to investigate. In *Raymond International, Inc.,*[25] the government allowed only two months for bid preparation. The Armed Services Board of Contract Appeals (ASBCA) found this to be an insufficient amount of time for the contractor to survey more than 311 kilometers of road in Thailand and prepare a bid. In *Pavement Specialists, Inc.,*[26] the contractor had twice attempted to investigate, but the government denied it access to the site each time. The board therefore excused the contractor's failure to investigate the site.

When the government supplies no information whatsoever about site conditions, the ASBCA tends to hold that the contractor is obliged to conduct a more detailed site investigation.[27] The rationale of these cases is that by not being provided any information, the contractor must realize that it has to gather its own information to ascertain site conditions.

Nevertheless, in this category of cases, at least when contractors conducted a reasonable site investigation without discovering the latent condition, the ASBCA has at times sustained differing site conditions claims because it finds that the government failed to disclose a latent condition of which the government was aware.[28] *Joseph A. Cairone, Inc.*[29] was an appeal by a contractor performing work on a site that proved to have subsurface water. The contractor experienced subsurface water problems and alleged that the government should have disclosed these problems, which had also been encountered by prior contractors on the same site. The board sustained the claim, stating that the government

---

[22] Tutor-Saliba, ASBCA No. 23,776, 79-2 B.C.A. (CCH) ¶ 14,137 (1979).

[23] ASBCA No. 29,581, 85-1 B.C.A. (CCH) ¶ 17,764 (1984).

[24] ASBCA No. 26,136, 83-2 B.C.A. (CCH) ¶ 16,612 (1983).

[25] ASBCA No. 13,121, 70-1 B.C.A. (CCH) ¶ 8,341 (1970).

[26] ASBCA No. 17,410, 73-2 B.C.A. (CCH) ¶ 10,082 (1973).

[27] *See, e.g.,* Commercial Mechanical Contractors, Inc., ASBCA No. 25,695, 83-2 B.C.A. (CCH) ¶ 16,768 (1983).

[28] *See, e.g.,* Leonard Blinderman Constr. Co., ASBCA No. 18,946, 75-1 B.C.A. (CCH) ¶ 11,018 (1975) (visual inspection did not reveal depression in structural support system and government did not disclose it); Edgar M. Williams, Gen. Contractor, ASBCA No. 16,058, 72-2 B.C.A. (CCH) ¶ 9,734 (1972) (double roofs are considered unusual although they are common to Fort Polk and government knew that).

[29] ASBCA No. 20,504, 81-2 B.C.A. (CCH) ¶ 15,220 (1981).

reasonably and in good conscience should have disclosed the problems encountered by prior contractors. A similar situation occurred in *Commercial Mechanical Contractors, Inc.,*[30] in which the ASBCA also found that the government should have disclosed subsurface water problems that other contractors had previously encountered at the site. The location of the site was over the Cache Creek Flood Plain.

---

[30] ASBCA No. 25,695, 83-2 B.C.A. (CCH) ¶ 16,768 (1983).

CHAPTER 6

COMMON LAW THEORIES FOR RECOVERY

Alan S. Pralgever

Alan S. Pralgever, a member of the litigation department of Brach, Eichler, Rosenberg, Silver, Bernstein, Hammer & Gladstone, has extensive experience in handling complex commercial and corporate litigation and has developed a special emphasis in construction law. He received his M.A. from Columbia University, New York, New York, and his J.D. from Antioch University in Yellow Springs, Ohio. Mr. Pralgever is a member of the New Jersey and New York bar associations and has been appointed to the Panel of Arbitrators for commercial and construction matters for the American Arbitration Association.

§ 6.1 Introduction

This chapter is designed to serve as a guide through a variety of viable noncontractual common law causes of action, defenses, and remedies that may be applied in differing site conditions cases. Such noncontractual theories as fraud, misrepresentation, and negligence are rooted in traditional tort law. When used effectively, these claims can play a significant role in persuading uncooperative defendants to settle valid differing site conditions claims. Frequently, noncontractual claims result in significantly larger recoveries than those rooted in contract. The clear and present danger of risking such loss is often just the impetus recalcitrant defendants need to convince them to resolve the underlying differing site conditions claims.

A typical fact scenario that gives rise to a noncontractual cause of action is when an owner or its representative induces a builder to submit an unrealistically low bid by misrepresenting the site conditions, omitting material facts regarding known site conditions, failing adequately to test the site to determine actual subsurface conditions, or failing to provide bidders with adequate time or opportunity to do so. Depending on the factual predicate and applicable federal or state law, the circumstances underlying the claim may give rise to an action sounding in fraud, misrepresentation, or breach of warranty. Such actions are generally justified when the owner has made affirmative misrepresentations that mislead the contracting party or has omitted material or key factors, and the general contractor or construction manager or subcontractor has affirmatively met its obligation to test the subsurface conditions.

A second scenario might involve a tract developer or builder who chooses a site that later proves to be unfit for safe, durable construction. For example, a building may be situated in a flood plain, in wetlands area, or on contaminated land, or the subsurface soil conditions may make the site prone to landslides. Subsurface springs that are carelessly filled in or not adequately drained or diverted may raise the subsurface water table and cause basement flooding. In such cases, the contractor may be forced to expend great quantities of time, money, and effort to overcome conditions not anticipated. In such circumstances, the developer and builder may even be held liable to a subsequent or remote purchaser when the once-latent site defect becomes apparent.

Noncontractual theories provide fertile ground for contractors' claims that otherwise would not be actionable under a breach-of-contract theory. This is especially true when a builder reasonably relies upon an owner's representations as to site conditions that later prove to be inaccurate, incomplete, or misleading. A contract disclaimer that allocates the risk of site condition uncertainty solely to the contractor is not controlling in the face of an owner's affirmative misstatements or omissions of known material facts, and the contractor demonstrates reasonable reliance thereon.[1]

Often, noncontractual claims give rise to liability that exceeds the scope of remedies available in an ordinary contract action and, hence, serve as a persuasive tool to convince an owner to compensate the contractor for unanticipated expenses incurred due to differing site conditions. For example, consequential and punitive damages are generally not available remedies in a breach-of-contract case, especially if the contract contains a liquidated damages clause.[2]

---

[1] P.T.&L. Constr. Co. v. State, 108 N.J. 539, 531 A.2d 1330 (1987); Ell-Dorer Contracting Co. v. State, 197 N.J. Super. 175, 484 A.2d 356 (App. Div. 1984); Golomore Assocs. v. New Jersey State Highway Auth., 173 N.J. Super. 55, 413 A.2d 361 (App. Div. 1980); Sasso Contracting Co. v. State, 173 N.J. Super. 486, 414 A.2d 603 (App. Div.), *cert. denied,* 85 N.J. 101, 425 A.2d 265 (1980).

[2] *See, e.g.,* CCM Corp. v. United States, 20 Cl. Ct. 649, 621 n.1 (1990). The *CCM Corp.* court disallowed a plaintiff-contractor's claim for lost profits that resulted when the contracting entity caused delays that deprived the plaintiff of opportunities to enter into other contracts.

Yet, if a court finds that nondisclosure of site conditions was motivated by oppression, fraud, or malice, this opens the door to a host of noncontractual claims and punitive damages.[3] Noncontractual theories provide a vehicle to overcome damage limitation provisions, thus enabling a party to recoup damages for lost profits, delay, and extra costs. Accordingly, noncontractual claims often engender settlement, especially when one party justifiably fears that common law remedies may dramatically exceed contractual remedies.

In the event the contract is silent with regard to site condition uncertainties and related contingencies, more potential liability may result under a noncontractual theory than would have resulted from an ordinary contract action. Therefore, it is crucially important that parties to a construction contract consider drafting strategies, site condition disclosures, and other steps necessary to allocate risks and to insulate themselves from possible liability on a noncontractual theory. For example, a liquidated damages clause can serve as a hedge against uncertainty when neither party is fully aware of the site conditions to be encountered. However, such provisions will not likely be enforced in the face of one party's fraudulent nondisclosure or material omission of information regarding differing site conditions.

Planning is as strategic an element in litigating successfully as it is in building. Therefore, it is important to methodically consider the impact of raising a variety of noncontractual claims before initiating litigation. This can best be accomplished by considering a full litany of noncontractual claims, many of which are contained in this chapter.

Of course, the law in each state or jurisdiction may differ dramatically. Before initiating any action based on fraud, misrepresentation, or other common law theory, a claimant should carefully review the law in the appropriate jurisdiction. This investigation should probe applicable cause-of-action elements, defenses, and remedies in the relevant jurisdiction, as well as consider any pertinent federal and state tort claims acts if claims are asserted against a government entity. Many states have also enacted statutes that provide for sovereign immunity from suit, which should also be investigated.

§ 6.2 Fraud

Elements of fraud. Common law fraud traditionally consists of a false statement of material fact that is intended to induce reliance and that is a substantial factor in causing another's justifiable reliance to that person's detriment.[4] An action in fraud is thus typically indicated when an owner, developer, or public

---

The court held that such lost profits are "consequential damages" and could therefore not be recovered in an ordinary breach-of-contract action.

[3] Village Dev. Co. v. Filice, 90 Nev. 305, 526 P.2d 83, 89 (1974).

[4] *See, e.g.,* Restatement (Second) of Torts § 525 (1965).

entity seeks to induce favorable construction bids based on knowingly false and misleading statements or material omissions as to site conditions.[5]

In *L.J. McNulty, Inc. v. Village of Newport,*[6] the plaintiff contractor submitted a bid for the construction of subsurface water and sewage works. The town hired two engineering companies to conduct soil borings and to develop a subsurface rock profile chart based upon the materials uncovered by the boring samples. The plaintiff contractor relied upon these rock profiles in submitting its construction bid and, accordingly, was awarded damages predicated on its reasonable reliance.

Fraudulent misrepresentations are not just limited to statements that are patently false; it is enough if the misrepresentation is intended to mislead. For example, in *City of Salinas v. Souza & McCue Construction Co.,*[7] the defendant city's engineer ordered test borings of a construction site that were designed to avoid problem areas—not an uncommon situation. The reports generated by such testing could be construed to be fraudulent and misleading. Specifically, the engineer directed test borings at "locations which avoided the area of the greatest unsettled conditions."[8] The result was that the contractor was unaware of quicksand at the construction site and the fact that site soils could not be compacted as required by the contract.[9] Even if the test boring results themselves were accurate, the tests were conducted in such a way as to produce misleading inferences as to the overall subsurface soil conditions.[10] The *Souza & McCue* court ruled that the city fraudulently misrepresented and actively concealed unfavorable site conditions that directly affected the cost of completed construction on a sewer line.[11]

Such misrepresentation is compounded when the misleading site borings are disclosed immediately before the submission of bids (also not atypical) so that potential bidders do not have an opportunity to conduct adequate independent investigations. Under such circumstances, bidders have no reasonable means or opportunity to discover unanticipated difficulties and are directly damaged by the false and misleading information they have been provided.

Silence, the omission of material fact, may also be actionable as fraud if it induces a contractor to enter a construction contract on unfavorable terms. However, this is generally true only if the nondisclosing site owner or party has an

---

[5] Souza & McCue Constr. Co. v. Superior Court, 57 Cal. 2d 508, 370 P.2d 338, 339, 20 Cal. Rptr. 634 (1962).

[6] 290 Minn. 117, 187 N.W.2d 616 (1971).

[7] 66 Cal. 2d 217, 424 P.2d 921, 923, 57 Cal. Rptr. 337 (1967).

[8] 424 P.2d at 923.

[9] *Id.* at 924.

[10] *Id.* at 923.

[11] *Id.* at 922–24.

affirmative duty to disclose.[12] Courts differ as to what circumstances will give rise to a duty to disclose. Generally, courts that impose such a duty do so at a relatively simple threshold, holding that if the nondisclosing party possesses any material information that would affect the bids of potential contractors, the information must be disclosed. An owner has the general duty "to impart its knowledge of the difficulties to be encountered in a project" and will bear the risk if the contractor is unable to perform according to contract provisions.[13]

In *P.T.& L. Construction Co. v. State,*[14] the New Jersey Supreme Court ruled that when the state actually makes false representations in its contract averments that are more than gratuitous and amount to positive averments of site conditions, it will remain liable to a public contractor despite a general exculpatory clause in the contract. The court went so far as to suggest that "[i]n some cases, actual concealment of information may be considered a false factual representation."[15] However, the court warned that inferential conclusions from contract documents may not be considered false representations in the face of sufficiently clear and unambiguous disclaimers of liability. From this we can conclude that false representation findings are largely fact dependent.

Similarly, in *Steinberg v. Fleisher,*[16] a contractor alleged that the plaintiff owners fraudulently induced it to enter into a construction contract by failing to disclose the existence of subsurface rock and soils that could not be easily compacted. The court concluded that the issue of whether or not a contractor had met its burden of proof on a fraud claim was a question of fact for a jury to decide.[17]

---

[12] Wiechmann Eng'rs v. State, 31 Cal. App. 3d 741, 107 Cal. Rptr. 529, 535 (1973). "Courts uniformly distinguish between the misleading half-truth, or partial disclosure, and the case in which defendant says nothing at all. The general rule is that silence alone is not actionable."

[13] City of Salinas v. Souza & McCue Constr. Co., 66 Cal. 2d 217, 424 P.2d 921, 923, 57 Cal. Rptr. 337 (1967). The California Supreme Court subsequently suggested that the test of a site owner's duty to disclose may be narrower that the broad duty set forth in *Souza & McCue.* Specifically, an action for fraudulent nondisclosure may arise when incomplete disclosures are likely to mislead, when only the defendant site owner has access to material information, and when the defendant actively conceals information from the plaintiff. Warner Constr. Corp. v. City of Los Angeles, 2 Cal. 3d 285, 466 P.2d 996, 1001, 85 Cal. Rptr. 444 (1970).

Similarly, a county building inspector bears the duty to disclose defective construction; this duty arises out of the inspector's "duty to perform his job with diligence and candor [which] is imposed upon him for the benefit of the public." Cooper v. Jevne, 56 Cal. App. 3d 860, 128 Cal. Rptr. 724, 732 (1976).

[14] 108 N.J. 539, 531 A.2d 1330 (1987).

[15] 108 N.J. at 562.

[16] 706 S.W.2d 901, 904–5 (Mo. Ct. App. 1986).

[17] *Id.* at 905. Specifically, the elements of a cause of action for fraudulently inducing the contractor to enter into a construction contract by failing to disclose the existence of subsurface soil conditions consisted of the following elements: "Whether representations were made, whether the representations were false, and whether defendants were entitled to rely on their representations." *Id.*

Other jurisdictions have taken a wholly opposite view, finding that contracting parties ordinarily owe each other no duty to disclose. Such jurisdictions follow narrower criteria in determining whether a site owner has a duty to divulge known site condition information.[18] For example, a duty to disclose can arise if information is peculiarly within the owner's knowledge and is not readily obtainable by the contractor, or if the contractor inquired into site conditions and was answered incompletely or untruthfully.[19] Alternatively, a court may impose a duty to disclose only when the public owner occupied "so uniquely-favored a position with regard to the information at issue that no ordinary bidder in the [contractor's] position could reasonably acquire that information without resort to the state."[20]

There are many more fraud-misrepresentation cases pertaining to public as opposed to private agents. This is because generally the government (federal, state, and local) is perceived to have an affirmative duty to disclose. However, this body of law is readily applied to private projects as well and provides a viable basis for initiating noncontractual claims. Unless the contractor has been given ample direction in the contract, as well as adequate time and opportunity to investigate subsurface conditions, allegations of fraudulent misrepresentation can give rise to a viable cause of action and enable a contractor to assure adequate recovery. These causes of action also can help to overcome or mitigate the effectiveness of no-damages-for-delay and liquidated damages provisions.

Because fraud claims are largely factually rooted, it is important to verify all facts carefully. A complaint that fails to allege fraud with sufficient particularity is readily subject to a motion to dismiss. This is true both in state and federal jurisdictions. For example, insufficient allegations of fraud are particularly susceptible to dismissal pursuant to Federal Rule of Civil Procedure 12(b). To plead fraud with sufficient particularity, a complaint must allege facts showing fraud, sources that support the alleged specific facts, and the basis from which inference of fraud may be fairly drawn. Otherwise, a motion to dismiss will be sustained.[21] In addition, Federal Rule of Civil Procedure 11 provides for stiff sanctions for bad-faith allegations.[22] Therefore, it is important to investigate the facts thoroughly before alleging fraud or misrepresentation.

---

[18] Simpson Timber Co. v. Palmberg Constr. Co., 377 F.2d 380, 385 (9th Cir. 1967).

[19] *Id.*

[20] Morrison-Knudsen Co. v. State, 519 P.2d 834, 841 (Alaska 1974).

[21] *In re* Frank B. Hall & Co., 693 F. Supp. 1460 (S.D.N.Y. 1988); *see also* Friedlander v. NIMS, 755 F.2d 810 (11th Cir. 1985); *In re* Midlantic Corp. Shareholder Litig., 755 F. Supp. 226 (D.N.J. 1990).

[22] Saltany v. Reagan, 886 F.2d 438 (D.C. Cir. 1989), *cert. denied,* 110 S. Ct. 2172 (1990) (rule 11 sanctions were imposed when attorney "surely knew" that complaint had no hope whatsoever of success). *See also* Thrush v. Morrison, 665 F. Supp. 372 (M.D. Pa. 1987), *aff'd,* 845 F.2d 1016 (3d Cir.), *cert. denied,* 109 S. Ct. 74 (1988); Sharjah Inv. Co. v. P.C. Telemart, Inc., 108 F.R.D. 90 (S.D.N.Y. 1985).

Defenses to fraud. A plaintiff's failure to meet stringent proof requirements will readily defeat claims of fraud. Even if the site owner discloses information that proves to be incomplete or inaccurate, such failures are often insufficient to give rise to an action for fraud. It is quite difficult to prove that an owner engaged in a deliberate and calculated attempt to conceal material information to deceive the bidder or to induce the bidder into submitting an unrealistically low bid.[23] Absent such proof of culpable intent to deceive, an action for fraud generally will not be sustained.[24] Obviously, demonstrating that an owner had a pecuniary interest in failing to disclose critical information goes a long way to proving intent. However, a litigant may have to engage in costly and time-consuming pretrial discovery to prove fraud.

Nondisclosure of site hazards may not be deemed fraudulent when conditions would be readily apparent in an on-site inspection. A contractor cannot realistically claim to have been fraudulently misled by site condition specifications that casual observation would have disproved.[25] However, a party can escape successful application of the "contributory negligence" defense by conducting a thorough on-site inspection and by directing inquiries as to site conditions to the owner or the site engineers.[26] In the alternative, contractors should take the time to take borings or otherwise test subsurface conditions on their own. If an owner appears reluctant to provide adequate time or access to the site to perform such tests, the contractor should be on notice that subsurface conditions may differ materially from those specified by the contract.

Similarly, a contractor is charged with full knowledge of site condition reports readily available for inspection for a reasonable length of time before submitting a bid or beginning work.[27] A claim of fraudulent misrepresentation will fail when the contractor, "although aware of their existence, made no effort to obtain the original boring logs" that would have reasonably represented the subsurface conditions to be encountered.[28] A contractor's negligent failure to ascertain soil conditions from readily available sources will bar a claim that the site owner fraudulently concealed information that a prudent bidder would have

---

[23] Wiechmann Eng'rs v. State, 31 Cal. App. 3d 741, 107 Cal. Rptr. 529, 535 (1973); Warner Constr. Corp. v. City of Los Angeles, 2 Cal. 3d 285, 466 P.2d 996, 1002, 85 Cal. Rptr. 444 (1970); Kaiser Indus. Corp. v. United States, 340 F.2d 322, 330–31 (Ct. Cl. 1965); Waggoner v. Midwestern Dev., Inc., 83 S.D. 57, 154 N.W.2d 803, 806–7 (1967).

[24] Kaiser Indus. Corp. v. United States, 340 F.2d 322, 331 (Ct. Cl. 1965).

[25] Wiechmann Eng'rs v. State, 31 Cal. App. 3d 741, 107 Cal. Rptr. 529, 535 (1973).

[26] Gogo v. Los Angeles County Flood Control Dist., 45 Cal. App. 2d 334, 114 P.2d 65, 68, *motion denied,* 47 Cal. App. 2d 96, 117 P.2d 392 (1941).

[27] Oregon State Highway Comm'n v. DeLong Corp., 9 Or. App. 550, 495 P.2d 1215, 1221 (1972), *cert. denied,* 411 U.S. 965 (1973); Wiechmann Eng'rs v. State, 31 Cal. App. 3d 741, 107 Cal. Rptr. 529, 533, 537 (1973).

[28] Oregon State Highway Comm'n v. DeLong Corp., 9 Or. App. 550, 495 P.2d 1215, 1221–22 (1972), *cert. denied,* 411 U.S. 965 (1973).

sought.[29] This duty to inspect and investigate available site descriptions and conditions also imposes upon the contractor constructive knowledge of excavation difficulties that "an experienced contractor" or "an experienced engineer" would "reasonably understand" to be described.[30]

Construction contracts typically contain disclaimer clauses stating that all potential bidders have a duty to investigate site conditions and place their bids in reliance upon their own independent investigations. Such disclaimers are not effective in the face of an owner's active and fraudulent concealment of actual site conditions.[31] Nor are such disclaimers enforced when contract specifications are "misleading half-truths" due to nondisclosure of information necessary to a complete and accurate understanding of the information actually disclosed.[32] Whether the purported fraud arises from active concealment or from willfully false statements, courts agree that "[i]t would be inequitable to permit [the owner who fraudulently induced a contractor to submit an unrealistically low bid] to enforce the literal terms of the contract.[33]

In sum, a contract disclaimer will not insulate an owner from liability arising out of deliberate fraud when the contractor did not have a reasonable opportunity to discover the true subsurface site conditions.

Remedies. Money damages for fraudulently misrepresented site conditions typically include components that reflect the added expense incurred in reasonable reliance upon the fraudulent misrepresentations. Such damages are typically measured by the difference between "the fair and reasonable cost of the actionable performance, and what it would have been in the absence of [fraudulent] misrepresentation."[34]

Reimbursement for indirect overhead expenses and reasonable compensation for lost profits on the additional work are also recoverable costs and expenses.[35] To recover compensation for lost profits and overhead expenses, the contractor is not required to prove its exact losses on the contract in question; it is enough for the claimant to prove overhead and profit percentages that are customary in the

---

[29] Wiechmann Eng'rs v. State, 31 Cal. App. 3d 741, 107 Cal. Rptr. 529, 536 (1973).

[30] Arthur A. Johnson Corp. v. Commonwealth, 318 Mass. 88, 60 N.E.2d 364, 369 (1945).

[31] City of Salinas v. Souza & McCue Constr. Co., 66 Cal. 2d 217, 424 P.2d 921, 923, 57 Cal. Rptr. 337 (1967).

[32] Warner Constr. Co. v. City of Los Angeles, 2 Cal. 3d 285, 466 P.2d 996, 1002, 85 Cal. Rptr. 444 (1970).

[33] Gogo v. Los Angeles County Flood Control Dist., 45 Cal. App. 2d 334, 114 P.2d 65, 69, *motion denied,* 47 Cal. App. 2d 96, 117 P.2d 392 (1941).

[34] City of Salinas v. Souza & McCue Constr. Co., 66 Cal. 2d 217, 424 P.2d 921, 923, 57 Cal. Rptr. 337, 341 (1967). *See also* Daniel Hamm Drayage Co. v. Waldinger Corp., 666 F.2d 1213, 1215–16 (8th Cir. 1981).

[35] City of Salinas v. Souza & McCue Constr. Co., 66 Cal. 2d 217, 424 P.2d 921, 925, 57 Cal. Rptr. 337, 341 (1967); Daniel Hamm Drayage Co. v. Waldinger Corp., 666 F.2d 1213, 1216 (8th Cir. 1981).

construction industry.[36] Loss of goodwill may also be a compensable element of damages.[37]

However, the claimant does have the burden to prove that its damages are not speculative. Damages must be proven with a "reasonable degree of certainty"; mere speculation as to a contractor's losses will not suffice.[38] Failure to assert damages with the requisite degree of certainty will reduce a damage award accordingly.[39] This is because nationally a trend has developed to discount "total cost" claims, especially when there is no nexus between the damages asserted and the actual costs and expenses incurred.[40] Although it may not be possible to quantify every fraud or misrepresentation claim with total specificity, an effort should be made to quantify as much as possible. When such claims are not readily quantifiable, reliance on industry standards and expert testimony is appropriate.

§ 6.3 Misrepresentation

Elements of misrepresentation. A cause of action for misrepresentation arises out of a factual scenario similar to that giving rise to an action for fraud. Typically, an owner provides plans or site specifications that prove to be inaccurate, thus causing the contractor unforeseen expense in performing the contracted construction. Yet, a misrepresentation claimant has a less onerous burden of proof than the fraud plaintiff; the former need not prove that the misrepresentation was motivated by an intent to deceive.

Misrepresentation consists of three essential prima facie elements: (1) an actionable representation or omission; (2) the contractor's reasonable reliance; and (3) resulting damage.[41] "The general rule is that where plans or specifications lead a public [or private] contractor reasonably to believe that conditions

---

[36] Daniel Hamm Drayage Co. v. Waldinger Corp., 666 F.2d 1213, 1215 (8th Cir. 1981).

[37] Warner Constr. Corp. v. City of Los Angeles, 2 Cal. 2d 285, 466 P.2d 996, 1007, 85 Cal. Rptr. 444 (1970).

[38] Daniel Hamm Drayage Co. v. Waldinger Corp., 666 F.2d 1213, 1216 (8th Cir. 1981).

[39] Warner Constr. Corp. v. City of Los Angeles, 2 Cal. 2d 285, 466 P.2d 996, 1007, 85 Cal. Rptr. 444 (1970).

[40] *See, e.g.*, W.F. Magann Corp. v. Diamond Mfg. Co., 580 F. Supp. 1299, 1316–17 (D.S.C. 1984), which discussed the total cost approach to damages. The *Magann* court awarded a subcontractor the reasonable value of services actually rendered under a quantum meruit theory. The award included components reflecting the cost of labor, equipment, materials, overhead, and profits of 15%.

[41] *But see* Acchione & Canuso, Inc. v. Commonwealth Dep't of Transp., 501 Pa. 337, 461 A.2d 765, 768 (1983). The court ruled that a cause of action for misrepresentation for site conditions must contain five elements: (1) a positive representation of site conditions (2) of material fact (3) of which the contractor has no reasonable opportunity to learn through independent investigation, (4) which statement proves to be false, (5) resulting in damage to the contractor.

represented therein do exist and may be relied upon in bidding, he is entitled to compensation for extra expense incurred as a result of the inaccuracy of those representations."[42] Actionable representations include both affirmative misrepresentations set forth in plans and specifications, as well as omissions where the law imposes a duty on the site owner to disclose known site conditions. Reasonable reliance is tested by weighing factors on a case-by-case basis.

The clearest cases of misrepresentation arise when the site conditions set forth in contract specifications are simply inaccurate. The source of the inaccuracy may be intentional, negligent, or merely honest error, and each of these types of misrepresentation may be actionable.[43] A misrepresentation at a minimum occurs when the contracting entity makes positive averments as to site conditions that it knew or should have known were misleading.[44]

Spirit Leveling Contractors v. United States[45] involved a typical misrepresentation factual predicate in which the contractor alleged that it reasonably relied upon site conditions described in the contract that were materially different from those that the contractor encountered during its actual performance. In *Spirit Leveling,* the plaintiff entered into a contract that required it to remove sediment and to restore dikes in an Arizona watershed protection program. The plaintiff alleged that the contracting entity submitted estimates that misrepresented the amount of sediment to be removed as well as the scope of dike restoration required.

This case also illustrates the elements of a typical differing site conditions claim. Specifically, the plaintiff contractor must prove that site conditions described in a contract were "materially different from those encountered during

[42] Nelson Constr. Co. v. Port of Bremerton, 20 Wash. App. 321, 582 P.2d 511, 515 (1978).

[43] *See* Restatement (Second) of Torts §§ 552 (negligent misrepresentation), 552C (innocent misrepresentation) (1965); Al Johnson Constr. Co. v. Missouri Pac. R.R. Co., 426 F. Supp. 639, 648 (E.D. Ark. 1976), *aff'd,* 553 F.2d 103 (8th Cir. 1977) (both parties "expected the actual conditions to be substantially as represented by the plans, specifications, boring logs and seismic survey furnished to the bidders"; when contractor reasonably relied on misrepresentations set forth in boring logs and did not have a reasonable opportunity to discover the presence of subsurface rock itself, the site owner bore the added expense incurred when the actual site conditions proved to be more perilous than those set forth in the contract specifications). *But see* Rocell Constr. Co. v. State, 208 Misc. 364, 141 N.Y.S.2d 463, 473 (Ct. Cl. 1955) (contractor assumed the risk of honest mistakes in test boring reports that did not disclose the location of rock formations); Foster Constr. C.A. v. United States, 435 F.2d 873, 881 (Ct. Cl. 1970) (to succeed on claim that public entity misrepresented site conditions, plaintiff must prove "[s]ome degree of Government culpability—either untruth or such error as the legal equivalent"; mere misstatement as to site conditions may not be sufficient to support action for misrepresentation); Acchione & Canuso, Inc. v. Commonwealth Dep't of Transp., 501 Pa. 337, 461 A.2d 765, 768 (1983) (false or misleading statement is actionable under theory of misrepresentation only if such misrepresentation is made through "either gross mistake or arbitrary action on the part of the [contracting] agency or its engineer").

[44] Grow Constr. Co. v. State, 56 A.D.2d 95, 391 N.Y.S.2d 726, 728 (1977).

[45] 19 Cl. Ct. 84 (1989).

performance."[46] In addition, the plaintiff must prove that it reasonably relied upon misrepresentations set forth in the contract and related documents, and that it suffered damages as result of the material misrepresentations. This means that the plaintiff contractor incurred excess costs that were "attributed entirely to the materially different subsurface conditions met at the site."[47] In evaluating a misrepresentation claim, the court must determine whether the contractor behaved in a reasonable and prudent manner. This includes a determination as to whether the conditions encountered were reasonably foreseeable based upon the information the contractor was provided at the time it submitted its bid. A contractor can succeed on such a cause of action only if it incurred additional expenses while operating in a reasonable and prudent manner, and only if the site difficulties encountered were not reasonably foreseeable at the time of the bidding.

The *Spirit Leveling* court rejected the notion that unanticipated quantities of material to be removed could support a differing site condition claim.[48] The court also rejected plaintiff's misrepresentation claim because a reasonable contractor in plaintiff's position knew or should have known that the site was prone to flooding. Accordingly, the contractor could not claim that it was misled or reasonably relied upon the contracting entity's representations about site conditions.

Nor could plaintiff recover damages caused by "acts of God," such as excessive rains and flooding that necessitated increased excavation and sediment removal. The court concluded that such climatic conditions could not support a differing site condition claim. Again, the court found that the contractor reasonably should have known that the entire purpose for the project was to alleviate a critical flooding problem. This constituted adequate warning about the nature of the job and hazards to be encountered.[49] Absent an express allocation of liabilities in a contract, neither party to the contract can be held liable to the other when damages are caused by an act of God.[50]

Portable Rock Production Co. v. United States[51] applies a misrepresentation analysis similar to that set forth in *Spirit Leveling*. The plaintiff in *Portable Rock* contracted to construct roughly two miles of forest access roads in the Willamette National Forest in Oregon. One of the roads was constructed within the flood plain of the Willamette River. The contractor rested its claim upon alleged misrepresentations of the contracting entity that made excavation and

---

[46] *Id.* at 94.

[47] *Id.*

[48] *Id.* at 94–95. This conclusion does not square with other holdings, which have permitted contractors to recover when excess quantities of subsurface materials must be removed from a construction site.

[49] *Id.* at 95.

[50] *Id.* at 96.

[51] 4 Cl. Ct. 495 (1984).

construction of the roads more difficult than anticipated. Work was suspended for roughly one month due to wet ground conditions. In addition, wet conditions slowed construction by causing equipment to settle in mud. Overall, the plaintiff estimated that soggy soil conditions reduced construction efficiency by 50 to 60 percent.[52]

The *Portable Rock* court rejected the plaintiff's allegation that the muddy site conditions differed materially from the conditions set forth in the contract. Instead, the court stated that the contractor could readily observe water at the construction site by examining the contract plans or by examining the site itself. The court also rejected the notion that the defendant had withheld essential information within its knowledge.[53] Accordingly, the court dismissed the plaintiff's complaint.

Another example of a misrepresentation cause of action is found in *Al Johnson Construction Co. v. Missouri Pacific Railroad Co.*,[54] in which the contracting entity provided bidders with plans, specifications, boring logs, and a seismic survey. Both parties to the contract expected that the actual subsurface site conditions would be substantially synonymous with those set forth in the documents. The court found that the contractor reasonably relied upon misrepresentations set forth in the boring logs and did not have a reasonable opportunity to discover the presence of subsurface rock itself. In this circumstance, the site owner bore the added expense that the contractor incurred when the actual site conditions proved to be more perilous than those set forth in the contract specifications.

Similarly, in *City of Columbia v. Paul N. Howard Co.*,[55] a contractor alleged that it relied to its detriment on a misrepresentation by the city site owner when the contractor submitted its bid for construction of a public sewer. The city provided bidders with specifications that included logs recording the results of test borings that had been conducted to investigate subsurface soil conditions. The contractor discovered in constructing the sewers that the subsurface soil conditions were substantially different than those indicated by the logs. As a result, the contractor incurred additional unanticipated expenses.

The district court entered judgment in favor of the city, ruling that a contract disclaimer stating that the test boring logs were "for reference only and not a part of the contract documents" meant that even though the test boring logs were inaccurate, the contractor could not recover for the extra work.[56]

The Eighth Circuit reversed, ruling instead that the test boring logs were an integral part of the contract itself.[57] The court further ruled that the contractor

---

[52] *Id.* at 499.

[53] *Id.* at 503–4.

[54] 426 F. Supp. 639, 648 (E.D. Ark. 1976).

[55] 707 F.2d 338, 339 (8th Cir.), *cert. denied,* 464 U.S. 893 (1983).

[56] 707 F.2d at 339.

[57] *Id.* at 339–40.

should be permitted to pursue its tort claim against the city, in which the contractor alleged that it relied upon the misrepresentations set forth in the test boring logs. Whether this reliance was reasonable was an issue that the Eighth Circuit left for the trial court to decide on remand. As legal support for the contractor's misrepresentation tort claim, the court quoted Missouri authority which held "where plans or specifications lead a public contractor reasonably to believe that conditions indicated therein exist and may be relied upon in making his bid, he will be entitled to compensation for extra work or expense made necessary by conditions being other than so represented."[58]

A contractor can recover an equitable adjustment in the original contract price only if site conditions differ materially from those set forth in the contract and the contractor reasonably relies upon such misrepresentations. In *P.J. Maffei Building Wrecking Corp. v. United States,*[59] the court ruled that only if the actual site conditions encountered are materially different from those specified in the contract will the court permit an equitable adjustment for the increased cost of encountering the changed conditions. There, the government invited contractors to submit bids for the demolition and removal of a government building. The request for proposals stated that the building materials would be the property of the contractor, and that, therefore, the salvage value of building materials on the site should be considered in calculating a bid.[60]

The contractor recovered approximately 20 percent less salvageable steel than it estimated it could salvage from the building's demolition. Thereafter, the contractor submitted a differing site conditions claim, seeking an equitable adjustment of the contract price.[61] The court rejected this claim, ruling that the government did not make any specific representations as to the quantity of steel in the building.

The court also rejected the contractor's claim that the government had misrepresented existing site conditions. The contractor formulated its estimate in reliance upon drawings of the building in question. The court ruled that the contractor's reliance on these drawings was not reasonable and could not support the contractor's claim of misrepresentation. The government did not guarantee the accuracy of the structural drawings and included prominent disclaimers that the drawings were for the information of bidders only. Further, the government expressly disclaimed a guarantee as to the quantity, quality, or accuracy of the drawings.[62]

In sum, the court ruled that the contractor assumed the risk that information in the government's structural drawings would prove inaccurate. The court

---

[58] *Id.* at 340–41 (quoting Clark v. City of Humansville, 348 S.W.2d 369, 370, 374 (Mo. Ct. App. 1961)).

[59] 732 F.2d 913, 916–17 (Fed. Cir. 1984).

[60] *Id.* at 914.

[61] *Id.* at 915.

[62] *Id.* at 919.

denied any equitable adjustment in the contract price because the contractor's reliance was unreasonable. In addition, the court rejected the contractor's claim that the government had more accurate information as to the amount of salvageable structural steel on the site that it withheld.[63]

A more difficult situation arises when the alleged misrepresentation is not a positive averment, but rather an omission of material fact or failure to disclose information known to the owner.[64] At common law, businesspersons conducting commercial transactions and dealing with others at arm's length were generally under no affirmative duty to disclose information at all. This rule has gradually been judicially modified in many jurisdictions. The modern trend is for courts to impose a duty to disclose site condition information known to the governmental or private owner, but not known to the contractor. It arises under two circumstances: (1) when information "peculiarly within the scope of [the owner's] knowledge and not readily obtainable" is willfully withheld; or (2) when the owner "failed to give a complete and truthful answer to a broad inquiry" as to site conditions.[65]

Although such factors giving rise to a duty to disclose—superior knowledge willfully withheld or evasive answers to a contractor's inquiries—are typical of those that courts around the country weigh, the duty to disclose is a judicially developed doctrine. Accordingly, jurisdictions differ as to the extent to which site condition information must be disclosed and also as to what factors they will consider. For example, some jurisdictions extend the duty to disclose to all material information that the state or other contracting party had in its possession, at least to the extent that such information was not within the fair reach of the contractor.[66] Although a contractor is ordinarily charged with knowledge of any site conditions that would be apparent from casual observation, the contractor's duty to investigate will not discharge the public entity's duty to disclose superior knowledge when casual investigation would not put the contractor's knowledge on parity with that of the government.[67]

For example, when the results of boring tests reveal unfavorable site conditions known only to the government entity and such information is not accessible to the contractor, the public entity has been held in breach of its duty to

---

[63] *Id.* at 918–19.

[64] *See, e.g.,* Grow Constr. Co. v. State, 56 A.D.2d 95, 391 N.Y.S.2d 726, 728 (1977). The state disclosed that a construction site's soils were composed of "silt and sand overlying a very compact layer of glacial till." Although full site disclosures were present in boring tests provided to contract bidders, the state omitted to warn of the presence of gravel and boulders in the notes that warned only of sand and silt. Accordingly, the court ruled that the plaintiff contractor stated a viable cause of action for misrepresentation because the state knew or should have known that its site condition omission would have been misleading. 391 N.Y.S.2d at 728.

[65] Nelson Constr. Co. v. Port of Bremerton, 20 Wash. App. 321, 582 P.2d 511, 515, *review denied,* 91 Wash. 2d 1002 (1978).

[66] Hersey Gravel Co. v. State Highway Dep't, 305 Mich. 333, 9 N.W.2d 567, 570 (1943).

[67] 9 N.W.2d at 570.

disclose and is liable for such misrepresentation.[68] In *Hersey Gravel Co. v. State Highway Department,*[69] the contractor's casual observation of site conditions did not reveal the extent of the subsoil site difficulties that ultimately interfered with completion of the construction contract. Therefore, the state was not discharged from its duty to disclose known unfavorable subsoil conditions.[70] Further, the owner's duty to disclose will not be satisfied when the language of the specifications would not put an ordinary contractor on notice of the conditions to be encountered, and when such information is not readily made known to the bidder or set forth clearly in the specifications in terms that can be readily understood.[71]

In *CCM Corp. v. United States,*[72] the plaintiff contractor agreed to remove an existing concrete walkway and to replace it with a new, insulated, waterproof walkway. The contractor encountered unexpected tar pitch in a waterproofing system in the old walkway.[73] The tar pitch was incompatible with the asphalt waterproofing system that the contractor intended to install. The contract itself required the plaintiff to use this asphalt waterproofing membrane.[74] The contractor was therefore forced to undertake an expensive decontamination of the tar at the site.

The contractor sought reimbursement for the fair and reasonable value of its cost to remove the tar pitch contamination. The plaintiff also sought damages for lost profits, which the court dismissed in an earlier summary judgment motion.[75]

The gravamen of the plaintiff contractor's legal theory was that it relied upon representations and omissions in the plans and specifications that the defendant provided. For the plaintiff to succeed on such a claim, the court ruled that it was plaintiff's burden to prove that "it encountered subsurface or latent physical conditions at the site that differed materially than those expressly or impliedly indicated in the contract."[76] It was the role of the court to decide whether a reasonable and prudent contractor would have foreseen the presence of the tar pitch contamination.

Although there was no express representation in the contract itself regarding subsurface conditions, the court concluded that the contract impliedly suggested and supported an inference that the contractor would encounter compatible asphalt conditions rather than incompatible tar. In addition, the contract neglected

---

[68] Valentini v. City of Adrian, 347 Mich. 530, 79 N.W.2d 885, 889 (1956).

[69] 305 Mich. 333, 9 N.W.2d 567 (1943).

[70] *Id.,* 9 N.W.2d at 570.

[71] Valentini v. City of Adrian, 347 Mich. 530, 79 N.W.2d 885, 889–90 (1956).

[72] 20 Cl. Ct. 649 (1990).

[73] *Id.* at 651–52.

[74] *Id.* at 652 n.8.

[75] *Id.* at 653.

[76] *Id.* at 655.

to specify that the contractor's duties would include site decontamination. This also gave rise to the inference that such decontamination would not be necessary. In sum, the court concluded that the contractor justifiably and reasonably relied upon implications in the contract specifications that extensive decontamination and site remediation would not be necessary.[77]

The *CCM* court awarded damages that were calculated by assessing the cost of performing the unexpected additional work. The plaintiff met its burden of producing "sufficient evidence to enable the court to make a fair and reasonable approximation."[78] This approximation did not rise to the level of quantifiable mathematical precision; yet, the contractor's damages calculation was well within the ballpark of two estimates that the defendant's own expert witnesses produced.[79]

The second element of a misrepresentation cause of action is reasonable reliance upon inaccurate representations. Again, courts weigh a variety of factors to determine whether reliance is deemed reasonable. For example, when unanticipated site conditions would have been readily apparent upon casual inspection, a contractor is unlikely to succeed in establishing reasonable reliance.

On the other hand, the contractor may reasonably rely on the information provided when the public body had relatively greater access to information over a longer period of time, but the contractor was required to bid within a tight time frame. For example, when state officials conducted boring tests over several years and thereby gained valuable information not otherwise available to contractors who submitted bids based on only a month or so of preparation, the court is more likely to hold that reliance on such information is reasonable. Similarly, when information is uniquely within the possession of the site owner, and an independent inspection by the contractor is not feasible, the contractor would be justified in relying upon site condition specifications set forth by the owner.[80]

In sum, the owner may not unreasonably withhold information in its possession from bidders, including expert studies or reports. Such nondisclosure provides fertile ground for bid protests as well as postperformance causes of action. Finally, a misrepresentation cause of action must include proof of damages caused by the misrepresentation.

Defenses. The simplest yet most fact sensitive of the misrepresentation defenses is a challenge to the facial sufficiency of a plaintiff's prima facie case. Thus, evidence that the defendant landowner did not have superior knowledge

---

[77] *Id.* at 656.

[78] CCM Corp. v. United States, 20 Cl. Ct. 649, 657 (1990).

[79] *Id.* at 657.

[80] Acchione & Canuso Corp. v. Commonwealth Dep't of Transp., 501 Pa. 337, 461 A.2d 765, 769 (1983).

that was withheld willfully or otherwise or that information provided was truthful can defeat a misrepresentation claim.[81] Similarly, an allegation of misrepresentation will fail if plaintiff fails to sustain its burden of proving a causal link between the misrepresentation and plaintiff's damages.[82]

The defense of an insufficient prima facie case is most persuasive when circumstances indicate either that no reliance occurred or that such reliance was unreasonable. For example, when a contractor was alerted to the possibility that boulders would make a dredging contract difficult to perform, a court ruled that it would have been reasonably prudent for the contractor to investigate the site conditions carefully.[83] A misrepresentation action will not succeed in the face of the contractor's failure to do so. Further, absent proof that the contractor actually relied upon representations and that but for nondisclosure of material information the bid would have been different, a misrepresentation claim will fail.[84] Thus, proof of damages is inextricably linked to demonstrating liability. Again, total cost claims are typically not persuasive. Claims that are thoroughly quantified and well documented are generally more persuasive and less susceptible to attack by way of partial summary judgment motions.

Exculpatory clauses. Construction contracts typically include clauses in which the state or site owner disclaims liability for extra expenses that may arise if site conditions are other than anticipated. For example, a disclaimer clause may state that the contractor agrees to inspect the site itself and to submit a bid based on personal observations of site conditions rather than merely relying on information that the site owner provides.[85] These clauses may even go so far as to disclaim liability for the affirmative misrepresentations of public employees. Disclaimer clauses may not, however, by enforced when a public employee has made affirmative statements misrepresenting the actual conditions to be encountered.

Several states enforce exculpatory clauses that relieve the state or other contracting party of liability for representations made as to site conditions only if there is additional evidence that "(1) inspection would have been unavailing to

---

[81] Nelson Constr. Co. v. Port of Bremerton, 20 Wash. App. 321, 582 P.2d 511, 515, *review denied,* 91 Wash. 2d 1002 (1978).

[82] Chemical Bank v. State, 64 A.D.2d 755, 406 N.Y.S.2d 633, 634, *appeal denied,* 45 N.Y.2d 712, 383 N.E.2d 563, 411 N.Y.S.2d 1025 (1978).

[83] Nelson Constr. Co. v. Port of Bremerton, 20 Wash. App. 321, 582 P.2d 511, 515, *review denied,* 91 Wash. 2d 1002 (1978). *See also* Chemical Bank v. State, 64 A.D.2d 755, 406 N.Y.S.2d 633, 634, *appeal denied,* 45 N.Y.2d 712, 383 N.E.2d 563, 411 N.Y.S.2d 1025 (1978). The court ruled that the plaintiff contractor did not reasonably rely on its interpretation of contract specifications in light of the fact that the specifications "clearly indicated that the amount of topsoil to be removed from any given area along the job site had not been determined," 406 N.Y.S.2d at 634.

[84] A.S. Wikstrom, Inc. v. State, 52 A.D. 658, 381 N.Y.S.2d 1010, 1012 (1976).

[85] E.H. Morrill Co. v. State, 65 Cal. 2d 787, 423 P.2d 551, 552–53, 56 Cal. Rptr. 479 (1967).

reveal the incorrectness of the representations, or (2) the representations were made in bad faith."[86]

The parties' relative access to information is an important factor in determining whether or not a contractor has fulfilled its duty to inspect the site conditions. For example, assuming the site owner possesses information obtained over years of experience or site testing, such superior knowledge cannot be imputed to a contractor who has but a few weeks to develop a bid. Accordingly, a contract disclaimer placing the burden of site inspections upon a contractor who has been afforded relatively little time and opportunity to investigate will generally not be enforced.[87]

Courts differ in their willingness to enforce contract disclaimers that place the entire risk of differing site conditions upon the contractor. For example, in *McDevitt Mechanical Contractors, Inc. v. United States,*[88] the court ruled that the contracting entity successfully disclaimed all responsibility for misrepresentations set forth in detailed shop drawings. The parties agreed that the shop drawings contained a discrepancy by misrepresenting the elevation of a stainless steel inlet pipe. At issue was whether a reasonable contractor would have understood that it was responsible for measuring site dimensions and elevations itself, or whether a contractor could reasonably rely on the defective drawings.

The court cited various contract clauses that allocated the risk associated with inaccurate shop drawings to the contractor. It was the contractor's duty to obtain exact dimensions and elevations and to plan its work accordingly. The court concluded that the contract clearly required the contractor to properly install the inlet pipe "despite any errors in the shop drawings."[89] The contracting entity had "effectively disclaimed all responsibility for the verification of dimensions shown on the field drawings."[90] No reasonable or prudent contractor would have relied on the inaccurate shop drawings. Because the contractor had both the opportunity and the obligation to verify the dimensions of site conditions, the contractor was estopped from claiming that site conditions were other than those represented by the contracting entity.[91]

The common law defense of "assumption of the risk" is often invoked to shift the burden of investigating site conditions on the contractor. This is especially true when site conditions are readily apparent and would easily have been discovered had the contractor exercised reasonable diligence in inspecting the site and formulating a bid based on reasonably foreseeable site hazards.

---

[86] Public Constructors, Inc. v. State, 55 A.D.2d 368, 390 N.Y.S.2d 481, 485 (1977) (citations omitted); Grow Constr. Co. v. State, 56 A.D.2d 95, 391 N.Y.S.2d 726, 728 (1977); P.T.&L. Constr. v. State, 108 N.J. 539, 562, 531 A.2d 1330 (1987).

[87] Public Constructors, Inc. v. State, 55 A.D.2d 368, 390 N.Y.S.2d 481, 485 (1977); Grow Constr. Co. v. State, 56 A.D.2d 95, 391 N.Y.S.2d 726, 728 (1977).

[88] 21 Cl. Ct. 616, 620 (1990).

[89] *Id.*

[90] *Id.*

[91] *Id.*

A site owner or public entity may be able to assert several of these viable defenses to a contractor's claim for equitable adjustment due to differing site conditions. In *Amino Brothers Co. v. United States,*[92] a contractor sought an equitable adjustment to its contract for the construction of a flood control project. Pursuant to the contract, the contractor built a low-water crossing across a river and used the crossing to haul away excavated material. A heavy rainfall resulted in uncontrolled flooding, which washed away the crossing. The court ruled that this flooding was an act of God and not a differing site condition. Thus, the contractor could not recover for the additional expenses incurred due to the washout. The crossing was subsequently washed out a second time when the government released water at an upstream dam to protect both the dam and the lives and property of those downstream. This time the court ruled that because the government was acting in its sovereign capacity to protect the public generally, the government was also not liable for damages caused by this second washout.[93]

These two defenses—an act of God and act of the sovereign—enabled the government to escape liability for an equitable adjustment of the contract price. In addition, the court rejected the contractor's claim that it was misled both by statements made and information withheld as to the risk of flooding at the construction site.[94] This was because the contractor was experienced in flood control and was on notice according to the terms of the specifications that the purpose of the construction was to control flooding. Plaintiff was also aware of the existence of an upstream flood control dam. Under these circumstances, it was reasonable for the contractor to anticipate flooding at the construction site.[95]

In sum, misrepresentation claims provide a viable basis for recovery when information about unanticipated subsurface site conditions has been wrongfully withheld and has had a significant effect on the cost and scope of the project. However, such claims generally fail when the resulting problems were within the contractor's knowledge and should have been anticipated.

Remedies. For a contractor that reasonably relies upon inaccurate or omitted site condition descriptions, the basic measure of damages is the amount of time or money necessarily incurred to deal with the conditions misrepresented.[96] When the actions of both parties contribute to the losses arising from misrepresentation, courts may equitably apportion damages to reflect the percentage of each party's responsibility in contributing to the additional costs of completion.[97]

---

[92] 372 F.2d 485, 490–91 (Ct. Cl. 1967).

[93] *Id.*

[94] *Id.* at 490.

[95] *Id.*

[96] E.H. Morrill Co. v. State, 65 Cal. 2d 787, 423 P.2d 551, 553, 56 Cal. Rptr. 479, 481 (1967).

[97] Grow Constr. Co. v. State, 56 A.D.2d 95, 391 N.Y.S.2d 726, 729 (1977).

In *Dawco Construction, Inc. v. United States,*[98] the United States Court of Appeals for the Federal Circuit ruled that damages can be calculated according to the jury verdict method only if three conditions are met:

1. That clear proof of injury exists
2. That there is no more reliable method for computing damages
3. That the evidence is sufficient for the court to make a fair and reasonable approximation of the damages.

Dawco involved a differing site claim in which the performance of a landscape portion of a housing refurbishment contract was hindered by site conditions that differed materially from those expected. Specifically, the landscaper encountered "unexpected massive rambling disbursed sub-surface running tree roots, rock, boulders, cobble, abandoned water lines not shown on the drawing, weed root masses, demolition debris, galvanized copper piping, asphalt, [and] concrete."[99] These conditions caused the landscaper to incur costs that otherwise would not have been necessary and that were not anticipated in the contractor's bid.

The *Dawco* court ruled that a contractor who wants to submit its damages claim to a jury bears the burden of proving that there is no more reliable method of calculating damages than the "guesstimate" of a jury verdict.[100] The court concluded that the contractor failed to meet its burden of proving that an injury existed, and that no more reliable method for computing damages than a jury verdict existed.[101]

In sum, a "total cost" estimate of damages in this case was simply too speculative. It was the contractor's burden to maintain records to sufficiently quantify what additional costs and expenses it actually incurred due to differing site conditions. The hazard of permitting a jury verdict under these circumstances was that unrealistic cost assumptions would result in unrealistic damage awards and would inequitably reward contractors who failed to prepare claims precisely with an unjustified windfall recovery.[102] Instead, the appropriate method to calculate damages is the "actual cost method," which requires the contractor to provide documentary evidence of increased expenses before an equitable adjustment for differing site conditions can be made.[103]

---

[98] 930 F.2d 872, 880 (Fed. Cir. 1991).

[99] *Id.* at 875 (citation omitted).

[100] *Id.* at 881.

[101] *Id.* at 880–81.

[102] *Id.* at 882.

[103] *Id.*

§ 6.4 Breach of Warranty

Prima facie elements. The theory that contract specifications and site plans carry with them an implied warranty of accuracy developed as an alternative to the tort theories of fraud and misrepresentation in which the sovereign or its agents were the misrepresenting parties.[104] Before the advent of state and federal tort claims acts, sovereigns were thought to be immune from liability for the tortious acts of their employees. Courts reasoned that a government entity's breach of warranty was more akin to an actionable breach of contract than to a nonactionable tort.[105] Accordingly, early remedies for breach of warranty mirrored the contract goal of making the nonbreaching party whole rather than punishing a wrongdoer for a fraudulent or tortious misrepresentation.[106] More recently, the breach-of-warranty theory of recovery has been substantially extended.

Two elements characterized early actions for breach of warranty: (1) erroneous statements in specifications that the land owner prepared and provided to contractors for use in developing a bid, and (2) the contractor's justifiable reliance on such misrepresentations.[107] For example, in *United States v. Atlantic Dredging,* the government conducted boring tests and invited contractors to bid for the job of dredging the Delaware River based on assertions that the material to be dredged was "mainly mud, or mud with admixture of sand."[108] In fact, the government knew but failed to disclose the presence of various impenetrable materials.[109] The Supreme Court ruled that the dredging company that submitted its bid was justified in relying upon this misrepresentation because the test boring results were stated with "assurance that it had a reliable foundation."[110]

Recent authorities follow this two-pronged approach in defining the elements of a breach-of-warranty theory. When an owner submits material representations as to site conditions upon which a contractor reasonably relies in submitting a bid, the owner bears the risk that its representations will prove to be inaccurate.[111] This includes assuming the risk for extra expenses incurred due to the fact that site conditions are not as represented.

---

104 *See, e.g.,* United States v. Atlantic Dredging Co., 253 U.S. 1, 40 S. Ct. 423, 425 (1920); United States v. Spearin, 248 U.S. 132, 39 S. Ct. 59, 61 (1918).

105 United States v. Atlantic Dredging Co., 253 U.S. 1, 40 S. Ct. 423, 425 (1920); Alpert v. Commonwealth, 357 Mass. 306, 258 N.E.2d 755, 764 (1970).

106 United States v. Atlantic Dredging Co., 253 U.S. 1, 40 S. Ct. 423, 425 (1920).

107 40 S. Ct. at 424–25.

108 *Id.* at 423.

109 *Id.* at 424.

110 *Id.* at 425.

111 Ray D. Lowder, Inc. v. North Carolina State Highway Comm'n, 26 N.C. App. 622, 217 S.E.2d 682, 692, *cert. denied,* 288 N.C. 393, 218 S.E.2d 467 (1975); J.A. Thompson & Son, Inc. v. State, 51 Haw. 529, 672, 465 P.2d 148, 153 (1970).

It is important to note that in some jurisdictions, neither fault nor negligence is a necessary element of a breach-of-warranty claim.[112] It is enough that the contract specifications are inaccurate and that the plaintiff relied upon the misrepresentations to its detriment.

Other courts have taken a less expansive view. These jurisdictions impose upon the contractor a duty of due care in evaluating the accuracy of the site owner's disclosures. A plaintiff builder cannot recover for extra work occasioned by misinformation unless he was "not reasonably able to discover the true facts for himself."[113] In addition, these jurisdictions require that misinformation will be actionable only if it involves a material inaccuracy, and not if it merely involves the mistaken judgment of the contractor plaintiff itself.[114]

Authorities differ as to whether a misrepresentation must be tainted by some wrongful motive to be actionable as a breach of warranty. Both willful nondisclosure accompanied by the intent to mislead[115] and honest misstatements that later prove to be inaccurate[116] supported warranty claims in early cases. The modern trend is to limit warranty liability to those circumstances in which the owner knowingly misrepresents or withholds material information within its knowledge.[117] If an owner's site condition representation is mistaken, or is accurate but leads to inferences that prove to be erroneous, modern courts are much more willing to let the risk in these situations rest with the contractor.[118]

A second modern limitation on the prima facie case for breach of warranty is that the owner impliedly warrants only that it has made a complete disclosure of such information that it has, not that such information will prove sufficient or complete.[119] Here again courts differ. Some extend the implied warranty to include an assurance that the site descriptions disclosed will be sufficient for the purpose of their disclosure, such as the contractor's use in developing a bid.[120] Other courts limit the implied warranty to assure only that material information known to the owner has not been withheld. In the second scenario, the risk that

---

[112] Waggoner v. Midwestern Dev., Inc., 83 S.D. 57, 154 N.W.2d 803, 807 (1967) (citing 77 C.J.S. *Sales* § 304 and W. Prosser, Prosser on Torts § 83, at 493 (2d ed.)).

[113] Jasper Constr. Co. v. Foothill Junior College Dist., 91 Cal. App. 3d 1, 153 Cal. Rptr. 767, 772 (1979) (citing Robert E. McKee, Inc. v. City of Atlanta, 414 F. Supp. 957, 959–60 (N.D. Ga. 1976)).

[114] Jasper Constr. Co. v. Foothill Junior College Dist., 91 Cal. App. 3d 1, 153 Cal. Rptr. 767, 772 (1979).

[115] United States v. Atlantic Dredging Co., 253 U.S. 1, 40 S. Ct. 423, 425 (1920).

[116] United States v. Spearin, 248 U.S. 132, 39 S. Ct. 59, 61 (1918).

[117] J.A. Thompson & Son, Inc. v. State, 51 Haw. 529, 465 P.2d 148, 155 (1970).

[118] 465 P.2d at 155.

[119] *See, e.g.*, Jasper Constr. Co. v. Foothill Junior College Dist., 91 Cal. App. 3d 1, 153 Cal. Rptr. 767, 772 (1979) ("there can be no liability of a public entity for extra work caused by plans and specifications that are merely incomplete").

[120] Alpert v. Commonwealth, 357 Mass. 306, 258 N.E.2d 755, 763 (1970).

the disclosed information will prove to be insufficient continues to rest with the contractor; the owner warrants only its truth.[121]

Other courts have ruled that an implied warranty arises not just as to the truth of subsurface conditions that the site owner discloses in its contract specifications, but also as to the feasibility of any mode of contract performance that the site owner requires as a term or condition of the contract itself. In *Midwest Dredging Co. v. McAninch Corp.*,[122] the court ruled that an implied warranty did not arise merely by the state's disclosure of such test boring results as the state possessed. Instead, the warranty arose when the state took the extra step of requiring that specific dredging techniques be employed, based on its subsurface soil profile tests.

Defenses. A contract clause that purports to disclaim responsibility for an owner's site condition representations may not be enforced when warranties as to site conditions and construction specifications are false and the contractor justifiably relies upon them.[123] A typical contract may require the contractor to examine the site, to review the owner's plans, and to assume the risk of any inaccuracy the plans contain.[124] Yet these disclaimers are not enforced in the face of inaccurate descriptions of site conditions or building specifications. In addition, contract disclaimers are not enforced if the contractor submits a bid in reliance upon the owner's disclosures, but does not have a feasible opportunity to investigate the site conditions for itself.[125]

Although a contract clause disclaiming liability for the accuracy of site specifications will not stand if the specifications prove to be inaccurate, such a contract disclaimer can be an important element of proof in defeating a warranty claim.[126]

At common law, contractors who agreed to construct a structure for a set price assumed the risk that site conditions would make performance more difficult than originally anticipated. Only if the unforeseen difficulties made performance totally impossible would a contractor be relieved of its duty to perform at the agreed contract price.[127] Despite this laissez-faire predisposition, courts in

[121] Jasper Constr. Co. v. Foothill Junior College Dist., 91 Cal. App. 3d 1, 153 Cal. Rptr. 767, 772 (1979).

[122] 424 N.W.2d 216, 222 (Iowa 1988).

[123] United States v. Spearin, 248 U.S. 132, 39 S. Ct. 59, 61 (1918).

[124] Al Johnson Constr. Co. v. United States, 854 F.2d 467, 468 (Fed. Cir. 1988).

[125] Midwest Dredging Co. v. McAninch Corp., 424 N.W.2d 216, 223 (Iowa 1988).

[126] Stuyvesant Dredging Co. v. United States, 834 F.2d 1576, 1581 (Fed. Cir. 1987). If the site owner specifically warns bidders that they must inspect both the site and available test borings to satisfy themselves as to the site conditions to be encountered, such a contract clause weakens a contractor's claim that it reasonably relied upon such information in drawing overbroad conclusions.

[127] United States v. Spearin, 248 U.S. 132, 39 S. Ct. 59, 61 (1918).

early breach-of-warranty cases were quick to hold that the contractor did not assume the risk that site conditions would differ materially from those set forth in the contract specifications, even if the misrepresentation was honest error. When the owner describes site conditions in contract specifications, it must assume the risk that site conditions may make performance more difficult than anticipated.[128]

By contrast, this risk of loss will rest on the contractor when the contractor fails to state a sufficient prima facie case. If the court finds that the owner's representations as to site conditions were not positive averments that materially affected performance, or that the contractor was not justified in relying upon the owner's representations, then no action for breach of warranty will lie.[129]

Further, owners have successfully defended against breach-of-warranty claims by proving that a contractor's bid reflected its assumption of a known risk. For example, the owner's test borings reveal site conditions only at the exact site of the drilling. If the contractor knew or should have known that the rock strata surrounding the drilling sites could differ materially from the actual drilling findings, then there is no implied warranty that all surrounding rocks would match the composition of the sites actually tested.[130] Thus, it is the bidder who assumes the risk that the site conditions surrounding test borings may differ materially from those that the drilling revealed. Any inference that the bidder makes as to surrounding site conditions it does at its own risk.[131]

However, not every description of site conditions stated in contract specifications rises to the level of a warranty upon which a contractor can reasonably rely. For example, a site owner who reports the average density of underwater materials that must be dredged does not warrant that the contractor will not encounter materials significantly denser than those set forth.[132]

The site owner can also defeat a warranty claim when the conditions complained of are readily visible in a casual inspection. In *Turnkey Enterprises Inc. v. United States,*[133] a contract provision that water would be available for the contractor's use from a nearby river was held not to constitute a warranty that the river contained water year round. The court further determined that plaintiff could not reasonably rely upon a Forest Service map from which plaintiff purportedly concluded that the river had a perennial flow sufficient for contract operations, inasmuch as the map had not been provided for the plaintiff as part

---

[128] 39 S. Ct. at 61.

[129] J.A. Thompson & Son, Inc. v. State, 51 Haw. 529, 465 P.2d 148, 153 (1970).

[130] 465 P.2d at 154.

[131] Warner Constr. Corp. v. City of Los Angeles, 2 Cal. 2d 285, 466 P.2d 996, 1000, 85 Cal. Rptr. 444 (1970).

[132] Stuyvesant Dredging Co. v. United States, 834 F.2d 1576, 1518 (Fed. Cir. 1987). A contractor who draws broad inferences from soil density data gathered from only a few specific spots does so at its own peril.

[133] 597 F.2d 750, 755 (Ct. Cl. 1979).

of the contract package and was not in any way referenced in the contract documents. In such a case, the contractor cannot successfully raise a breach-of-warranty claim if the condition could have been noted by simple observation.[134]

Warranty claims are also defeated when it is readily apparent that all parties had equal access to information. When site tests are fully disclosed and are accurate to the extent disclosed, then no warranty has been breached.[135] Conversely, when the owner withholds or misrepresents material information within its knowledge, then a breach-of-warranty claim is indicated.[136]

Although a contract clause standing alone that disclaims responsibility for site conditions cannot generally defeat a warranty claim, such a disclaimer can be effective if accompanied by the owner's complete disclosure of known information coupled with the parties' equal access to information.[137]

A contractor's failure to comply with the specifications set forth in a contract may defeat a warranty claim.[138] The United States Court of Appeals for the Federal Circuit recently ruled in *Stuyvesant Dredging Co. v. United States*[139] that a "design specification" carries with it an implied warranty that if the specifications are followed, an acceptable result will follow. However, it is important to note the distinction between a "design specification" and a "performance specification" in the context of an implied warranty cause of action. A *design specification* contains explicit requirements as to how the construction is to be performed. By contrast, *performance specifications* enumerate only required results and leave the methodology to achieve such results to the contractor. The *Stuyvesant* court ruled that the government's statements as to the soil conditions on a dredging site were performance specifications and were therefore not warranted to be accurate or adequate.[140]

In *Stuyvesant,* the government issued an invitation for bids in which it characterized the nature and density of materials to be dredged at the contract site. In two earlier bidding situations, the plaintiff contractor submitted bids only after conducting substantial preliminary investigations. Specifically, the contractor reviewed Army Corps of Engineers' records regarding previous dredgings of the channel in question. In addition, the contractor itself took physical samples of the materials to be dredged and conducted echo soundings of the channel floor.[141] The contractor submitted a bid on a subsequent dredging contract without conducting the above investigations. It did not review previous Army Corps

---

134 Jasper Constr. Co. v. Foothill Junior College Dist., 91 Cal. App. 3d 1, 153 Cal. Rptr. 767, 772 (1979).

135 J.A. Thompson & Son, Inc. v. State, 51 Haw. 529, 465 P.2d 148, 155 (1970).

136 465 P.2d at 155.

137 *Id.;* Wunderlich v. State, 65 Cal. 2d 777, 423 P.2d 545, 550, 56 Cal. Rptr. 473 (1967).

138 Al Johnson Constr. Co. v. United States, 854 F.2d 467, 470 (Fed. Cir. 1988).

139 834 F.2d 1576, 1582 (Fed. Cir. 1987).

140 *Id.*

141 *Id.* at 1579.

dredgings, did not take its own samples of the materials to be dredged, and did not conduct echo soundings of the channel floor. The contractor subsequently encountered site materials that were significantly denser and therefore more difficult to dredge than expected. This resulted in 24 days of unanticipated delay and required the removal of thousands more cubic yards of dredged material than expected.[142]

The court rejected the contractor's claim for an equitable adjustment of the contract price due to these differing site conditions because the technical provisions set forth in the contract did not warrant the density of the material to be removed. Instead, the contract warned only that the density readings provided were estimates of the range of densities that might be encountered. They were not representations of known maximum and minimum densities. In addition, the contract contained a disclaimer stating that bidders "are expected to examine the site of the work and the records of previous dredging," and to determine the site conditions for themselves and at their own risk.[143] In light of these contractual limitations, the court ruled that there was no misrepresentation upon which the contractor could reasonably rely. The court construed the contract specifications to be estimates, not warranties.[144]

Remedies. Because courts think of the breach-of-warranty theory as analogous to a breach of contract, its remedies parallel the compensatory goal typical of contract actions rather than the punitive nature of tort damages.[145] The goal in remedying a breach of warranty is to compensate the contractor for the extra work and expense incurred in performance over what would have been incurred had the site conditions been as represented in contract plans and specifications. Such recovery includes delay damages, home and field office costs, and impact damages, as well as the actual cost of the increased work.

§ 6.5 Negligence

Negligence causes of action in the differing site conditions context typically involve fact scenarios in which a builder or site design professional fails to exercise that degree of expertise that is usual for similar professionals in the same or similar circumstances.

A common law negligence cause of action has four prima facie elements:

1. A duty or obligation to exercise reasonable care under the circumstances to protect others against unreasonable and foreseeable risks

---

[142] *Id.* at 1580.

[143] *Id.* at 1581.

[144] *Id.*

[145] United States v. Atlantic Dredging Co., 253 U.S. 1, 40 S. Ct. 423, 425 (1920).

2. A breach of the duty to conform to the standard required
3. A close causal connection between the actor's conduct and the resulting injury
4. Resulting damages.[146]

For example, a builder must exercise a reasonable degree of care to protect the economic interests of the ultimate purchaser of the structure that it builds. This includes a builder's duty to exercise due care in selecting a suitable site for construction. In *ABC Builders, Inc. v. Phillips,*[147] the defendant builder breached this duty by constructing plaintiff's home on an unstable hillside site. Due to the presence of subsurface clay that was stable when dry but became slippery when wet, heavy rains resulted in a landslide that damaged plaintiff's home.[148] The developer had access to subsurface core drilling logs that would have revealed this problem. In addition, the nature of the subsurface material should have alerted the builder that the subsurface soils were not stable. In light of those two warning circumstances, the court concluded that a reasonably prudent builder would have conducted an investigation into the subsurface conditions that would have disclosed the subsurface soil instability.[149] It was negligent for the builder to select a site "in the path of an earth slide liable to occur at any time."[150]

Similarly, the California Supreme Court, in *Sabella v. Wisler,*[151] ruled that a builder would be liable under a negligence theory for constructing a house on insufficiently compacted soil, which resulted in the foundation and walls cracking. When the site settled, the floors were no longer level, and windows and doors would no longer open or close freely. Unlike the builder in *ABC Builders,* who had access to subsurface boring test results, the builder in *Sabella* failed to undertake any investigation at all as to the subsurface conditions. Had the builder engaged a competent engineer to conduct subsurface soil tests, it would have readily discovered that the site soil was not compacted and was unfit for construction.[152]

The *Sabella* court examined a number of factors that courts will weigh in evaluating a builder negligence claim. These factors include "the extent to which the transaction was intended to affect the plaintiff, the foreseeability of harm to him, the degree of certainty that he suffered injury, the closeness of the connection between the defendant's conduct and the injury suffered, and the policy of

---

146 W. Prosser, The Law of Torts § 30, at 143 (4th ed. 1971). *See also* ABC Builders, Inc. v. Phillips, 632 P.2d 925, 930–31 (Wyo. 1981).

147 632 P.2d 925, 935–38 (Wyo. 1981).

148 *Id.* at 938.

149 *Id.*

150 *Id.*

151 59 Cal. 2d 21, 377 P.2d 889, 27 Cal. Rptr. 689 (1963).

152 377 P.2d at 891.

preventing future harm."[153] The court concluded that defendant's decision to construct a foundation upon loose fill was a breach of defendant's duty to exercise reasonable care to prevent harm to the house's foreseeable future owners and was closely connected to the eventual damage that the house suffered.[154]

In *Conolley v. Bull,*[155] the court applied the *Sabella* factors in ruling that a builder would be liable for negligently constructing a house on an unsafe site following a landslide that damaged plaintiffs' house. The site was located on a slope over inadequately compacted soil and contained an underground spring for which defendant failed to provide drainage. Had defendant conducted a subsurface soil test, both the underground spring and the unstable subsurface conditions would have been discovered.[156] In addition, defendant would have been able to foresee that it would be necessary to construct a drainage system to channel the spring water and thereby avert a landslide. In sum, the California court ruled that the interests of preventing future harm to homeowners would best be served by imposing upon builders a duty to investigate subsurface conditions, especially when constructing on hillside lots.[157]

Owners, developers, surveyors, and site engineers may also be liable for the selection of an unsafe and ill-suited building site.[158] In *Village Development Co. v. Filice,*[159] plaintiffs purchased a vacant building lot from the defendant developer that included land in a flood plain that reasonably could be expected to flood once in five years. Although officers of the defendant developer were aware of this risk, defendant did not warn plaintiffs as to which areas of the site were off the flood plain and could safely support housing construction. Further, defendant's employee, who was a licensed land surveyor and engineer whose job it was to approve construction plans for all of the lots within the development, approved plaintiffs' proposed construction even though he knew or should have known the building site to be within the flood plain.[160] The court concluded that

---

153 *Id.* at 893 (quoting Stewart v. Cox, 55 Cal. 2d 857, 860, 362 P.2d 345, 13 Cal. Rptr. 521 (1961)).

154 *Id.*

155 258 Cal. App. 2d 183, 65 Cal. Rptr. 689, 697 (1968).

156 65 Cal. Rptr. at 689. *Cf.* Waggoner v. Midwestern Dev., Inc., 83 S.D. 57, 154 N.W.2d 803, 805–6 (1967). There, the court held that a developer who had filled in springs with dirt and rock was not negligent in doing so, even though this elevated the water table and was a factor in causing basement water seepage in plaintiffs' home. The court found that defendant did not know of the springs and that a reasonably prudent builder would not have been aware of the site problem. *Id.* at 806. Instead, the court held that defendant could be held liable under the theory of breach of an implied warranty of workmanlike construction. *Id.* at 809.

157 Conolley v. Bull, 258 Cal. App. 2d 183, 65 Cal. Rptr. 689, 698 (1968).

158 *See, e.g.,* Chemical Bank v. State, 64 A.D.2d 755, 406 N.Y.S.2d 633, 634, *appeal denied,* 45 N.Y.2d 712, 383 N.E.2d 563, 411 N.Y.S.2d 1025 (1978). The court found that the state could be held liable on a negligence theory because state employees failed initially to locate rough areas in a newly completed highway that needed to be ground smooth.

159 90 Nev. 305, 526 P.2d 83, 85 (1974).

160 526 P.2d at 85–86.

defendant breached its duty to warn plaintiffs of a concealed danger that they could not readily discover for themselves.[161]

Negligence defenses. Not every surveyor error in assessing site conditions will rise to the level of actionable negligence.[162] In *McKinnon v. Batte,*[163] the plaintiff property owners alleged that the defendant surveyor negligently failed to warn that the home plaintiffs purchased was located in a flood zone. After plaintiffs' uninsured home was damaged in a flood, plaintiffs filed suit alleging that the surveyor's failure to warn of the flood risk proximately caused their losses. Although the court conceded that a surveyor may be held liable for negligence when he fails "to exercise that degree of care which a surveyor of ordinary skill and prudence would exercise under similar circumstances," the court affirmed a jury verdict that the surveyor had in fact exercised that degree of skill typical of others in the locality.

The *McKinnon* court further noted that the plaintiffs failed to prove that they "detrimentally relied" upon the defendant's survey in deciding not to purchase flood insurance.[164] Instead, plaintiffs relied upon their informal conversations with neighbors, who had never suffered flood damages. Accordingly, plaintiffs failed to establish the third element of their prima facie case: they failed to prove a sufficiently close causal link between the alleged negligence of the defendant and their financial loss.[165]

Similarly, an architect will be liable for negligently designing a building only if its conduct falls below the standard of expertise of similarly situated professionals in its community. An architect will not be liable for structural defects that result from a failure to investigate subsurface site conditions of which other architects would not reasonably have been aware. In *Seiler v. Ostarly,*[166] the defendant architect designed and constructed an apartment building without discovering that the site soils were expansive clay that would swell when wet and contract when dry. This heaving and settling resulted in substantial damage to the foundation of plaintiff's building.

The architect did not conduct soil borings, which would have revealed the presence of the expansive clay, and would have indicated that pilings would be necessary in constructing a foundation. Instead, the architect questioned other

---

[161] *Id.* at 86.

[162] *See, e.g.,* Public Constructors, Inc. v. State, 55 A.D.2d 368, 390 N.Y.S.2d 481, 484 (1977). Subsurface test borings were compiled by both "skilled" and "unskilled" field workers. The state was not held liable for disclosing the work product of the unskilled field workers; instead, the gravamen of the successful cause of action rested upon the state's liability for its failure to disclose the subsurface test borings of its skilled workers. It was this nondisclosure that was actionable as misrepresentation.

[163] 485 So. 2d 295, 298–99 (Miss. 1986).

[164] *Id.* at 299.

[165] *Id.* at 298–99.

[166] 525 So. 2d 1207, 1208 (La. Ct. App. 1988).

architects and the local building inspector as to whether pilings were used in the surrounding areas generally. The court concluded that the failure to conduct soil borings was at worst a mere error of judgment for which negligence liability could not attach. The court thus determined that the defendant acted with the same degree of skill and care that was customary in defendant's profession and locale.[167]

Some states recognize a builder, developer, or design professional's contract privity defense to an allegation of negligence. The rule in such jurisdictions is that "the owner of a construction project may not recover compensation for economic damages caused by the negligence of an architect or engineer with whom it is not in privity of contract."[168]

Other jurisdictions follow the modern trend of disallowing the privity defense when the harm was to a foreseeable, intended beneficiary of the design or other professional's work. As a California court noted, the gravamen of an action for a professional's negligence is the tort of professional malpractice rather than breach of contract or product liability; the privity defense in such jurisdictions may be limited to the latter two actions.[169] Increasingly, design professionals may be held liable for their negligence in failing to uncover critical subsurface conditions.

In *Henshaw v. Edward E. Clark Engineers-Scientists, Inc.*,[170] a Florida court held that the privity defense was not available to the defendant engineering company that conducted a site safety inspection at the property of plaintiffs' employer. Defendant contracted with plaintiffs' employer to sample ground water and soil, to test for contamination, and to recommend cleanup and pollution control procedures. After conducting sampling tests and site remediation, defendant certified the employer's site to be relatively free of industrial pollutants. These safety conclusions later proved to be erroneous. Plaintiffs sued, alleging defendant had conspired with their employer to withhold information regarding the site's toxic contamination. The court concluded that plaintiffs stated a cause of action in negligence, and that defendant's duty of care extended to such reasonably foreseeable contract beneficiaries as employees who could be expected to work at the site and to be affected by toxic chemicals there.[171]

---

[167] *Id.* at 1209. *Compare* Grow Constr. Co. v. State, 56 A.D.2d 95, 391 N.Y.S.2d 726, 728–29 (1977). The court ruled that the state would be liable for damages incurred when its bridge designs prepared for a highway construction contract were negligently drawn. Specifically, the state invited bids based on proposal disclosures that stated that site soils were silt, sand, and compact glacial till; in fact, the soils proved to be water permeable, which resulted in unanticipated pumping of dewater cofferdams. In addition, the bridge plans misrepresented the height of rock elevations, which resulted in delay damages during the time that plaintiffs awaited revised and accurate bridge plans.

[168] Key Int'l Mfg., Inc. v. Morse/Diesel, Inc., 142 A.D.2d 448, 536 N.Y.S.2d 792, 793 (1988).

[169] Cooper v. Jevne, 56 Cal. App. 3d 860, 128 Cal. Rptr. 724, 728 (1976).

[170] 490 So. 2d 161, 163 (Fla. Dist. Ct. App. 1986).

[171] *Id.*

An additional successful defense to a negligence action arises when there is no adequate proof that the defendant knew or should have known of the site conditions alleged to have caused plaintiff's harm.[172] For instance, a contractor's failure to foresee troublesome site conditions from plain indications on plans, specifications, or other documents can result in the denial of the contractor's claim for an equitable adjustment in the contract price. In *Granite-Groves v. Washington Metropolitan Transit Authority,*[173] a contractor encountered unanticipated sand and water intrusion problems while constructing part of the Washington, D.C. subway system. Based upon pre-bid soil reports, boring logs, and site profiles, the contractor expected that tunneling conditions at one of two disputed sites would be "easy and relatively dry."[174] Instead, the contractor encountered sandy material that permitted water to flow into the excavation. The court concluded that the site condition reports provided to the contractor "contained clear indications of the problematic subsurface conditions," and that "a reasonable contractor would have foreseen the unfavorable mining conditions encountered" at that site.[175]

By contrast, the *Granite-Groves* court ruled that the contractor could not reasonably foresee unfavorable conditions at a second construction site. This site had relatively sandy material that permitted seepage in-flow. Yet the court concluded that boring logs and soil profiles did not alert the contractor to the presence of unfavorable tunneling conditions at this second site.[176]

The *Granite-Groves* court awarded an equitable adjustment in the contract price because the subsurface conditions that the construction contractor encountered "differed materially from those indicated in the contract documents."[177] The court ruled that the contractor could recover only its costs incurred that were directly attributable to the presence of sand and water in-flow at the second construction site. It was the contractor's burden to prove precisely the amount of equitable adjustment due to these specific causes.[178]

§ 6.6 Other Equitable Doctrines

Quasi-contract. A quasi-contract is technically not a contract at all. Instead, it is a contract implied by law in situations in which it would be unfair to allow one party to keep the benefit conferred upon it by another without paying the

[172] Waggoner v. Midwestern Dev., Inc., 83 S.D. 57, 154 N.W.2d 803, 806 (1967).

[173] 845 F.2d 330, 331 (D.C. Cir. 1988).

[174] *Id.* at 338–39.

[175] *Id.* at 338–40.

[176] *Id.* at 340–41.

[177] *Id.* at 342.

[178] *Id.*

value of the benefit bestowed. The goal of quasi-contract is to prevent the unjust enrichment of the benefited party.[179]

Unjust enrichment. As the term *unjust enrichment* suggests, some fact or circumstance must make it fundamentally unfair for the benefit to remain with the defendant. "In the absence of fraud or bad faith, a person is not entitled to compensation on the ground of unjust enrichment if he received from the other that which it was agreed between them the other should give in return."[180]

Quantum meruit. In *W.F. Magann Corp. v. Diamond Manufacturing Co.,*[181] the court employed the doctrine of quantum meruit in awarding damages to a subcontractor due to extra unanticipated expenses incurred when the subcontractor encountered differing site conditions. The defendant subcontractor agreed to dredge channels and a deposition basin that would result in a navigable entrance to Murrell's Inlet, South Carolina.[182] Under the terms of the contract, the subcontractor "accepts the sole responsibility for unanticipated costs," including "topography and ground surface conditions; nature and quantity of surface materials to be encountered; [and] subsurface conditions."[183]

Testimony at trial proved that unexpected site conditions caused the subcontractor unanticipated difficulty in performing the dredging. Sand fell into the dredged channel; the side slopes of the channel began to fall; and shoaling was an extensive problem. The result of these conditions was to require the dredging of considerably more material than the subcontractor originally anticipated.

The court concluded that representations set forth in the contract specifications as to the nature and characteristics of the sand to be dredged were inaccurate.[184] Whereas the contract specifications stated that the dredging work would be standard and typical, in fact the sands were unstable and were susceptible to erosion.[185] The court held that the subcontractor was entitled to recover quantum meruit damages for the "reasonable value of its materials, equipment used and services."[186] The subcontractor was also entitled to rescind the contract in addition to seeking restitution.[187]

---

[179] *See, e.g.,* Callano v. Oakwood Park Homes Corp., 91 N.J. Super. 105, 108–9, 219 A.2d 332 (1966).

[180] S&M Constructors, Inc. v. City of Columbus, 70 Ohio St. 2d 69, 434 N.E.2d 1349, 1351 (1982).

[181] 580 F. Supp. 1299 (D.S.C. 1984).

[182] *Id.* at 1301–2.

[183] *Id.* at 1302.

[184] *Id.* at 1306.

[185] *Id.*

[186] *Id.* at 1315.

[187] W.F. Magann Corp. v. Diamond Mfg. Co., 580 F. Supp. 1299, 1315 (D.S.C. 1984). The *Magann* court's decision also contains a useful discussion of contractors' claims and counterclaims under the Miller Act, 40 U.S.C. § 270a–270d (1988). 580 F. Supp. at 1315–17.

The court awarded quantum meruit damages according to the "total cost" approach. This meant that the subcontractor was awarded restitution for labor and materials, general administrative overhead, and profits at a rate of 15 percent.[188]

Similarly, in *Steinberg v. Fleisher,*[189] a contractor asserted a quantum meruit claim to recover for extra work that was not within the contemplation of the parties at the time they entered into the contract. The contractor incurred extra costs when it encountered unexpected rock that had to be blasted, and subsurface soils that could not be compacted as expected. The court permitted the contractor to submit both its quantum meruit theory and its breach of contract theory to the jury. This did not result in a duplication of damages because the court limited the quantum meruit claim to extra work above and beyond that required by the contract. Extra work is "not controlled by the contract."[190] It was for the contractor to prove the reasonable value of its additional expenses incurred as a result of the site owners' misrepresentation of subsoil conditions.[191]

Mutual mistake. Often, neither the contractor nor site owner is aware of the full extent of construction difficulties that will arise due to site conditions. Whether such a mutual mistake of material fact will give rise to an action for relief depends on whether either party took definitive steps to assume the risk or to induce the other's reliance on its own mistaken interpretation.

For example, in *Hollerbach v. United States,*[192] the Court ruled that it was the government, not the contractor, that held apparently superior knowledge when it provided specifications for a dam repair job. Although the specifications stated that the dam was backed with "broken stone, sawdust, and sediment," in fact the backing consisted of "soft, slushy sediment" that required additional, unanticipated excavation.[193] In light of the government's positive averment regarding site conditions, the Court held that the loss occasioned by this mutual mistake should rest with the government.[194] A contract clause disclaiming the government's liability for unexpected conditions and requiring contract bidders to conduct their own site inspections was held ineffective to shift the risk of mistake.

---

188 W.F. Magann Corp. v. Diamond Mfg. Co., 580 F. Supp. 1299, 1315–16, 1318 (D.S.C. 1984).

189 706 S.W.2d 901, 906–7 (Mo. Ct. App. 1986).

190 *Id.* at 906.

191 *Id.* at 907.

192 233 U.S. 165, 34 S. Ct. 553, 556 (1914).

193 34 S. Ct. at 554.

194 *Id.* at 556.

CHAPTER 7

CLAIM PREPARATION: PRACTICAL GUIDE

Eugene Schaffel
Brian P. Craig

Eugene Schaffel is a partner in the law firm of Buckley, Treacy, Schaffel, Mackey and Abbate. He is a graduate of the College of the City of New York and New York University School of Law. He has practiced construction law for over 30 years, and has served in various town or village positions in Suffern and Ramapo, New York.

Brian P. Craig is an associate with the law firm of Buckley, Treacy, Schaffel, Mackey & Abbate. Mr. Craig earned an undergraduate degree in history and government from Daemen College in Amherst, New York. He graduated from the Thomas M. Cooley Law School in Lansing, Michigan, where he served as editor-in-chief of the *Law Review*. Mr. Craig is a member of the New York, New Jersey, and District of Columbia Bars.

§ 7.1 Introduction

Although the majority of construction contracts may contain a differing site conditions clause under which the owner shares in the risk of encountering a changed condition, the trend seems to be to shift that burden to the contractor. More often than not, both public and private construction contracts contain exculpatory clauses that attempt to relieve the owner of all responsibility for unforeseen differing site conditions.

Whether a contract contains a changed conditions clause will undoubtedly have a profound effect on the owner's cost of construction as well as a contractor's ability to turn a profit. Without a differing site conditions clause, the contractor may be forced to increase its bid by including a contingency item to safeguard against the possibility of encountering an unexpected condition. The result is an overall increase in the owner's total cost of financing the construction. If, however, the contractor elects not to include such a contingency item, the contractor alone bears the risk of the changed condition which, if encountered, could be devastating to its profitability and perhaps even its survival.

A changed condition will most likely prevent the contractor from completing the project in a timely manner and subject it to the consequences that result therefrom (such as liquidated damages and/or engineering charges). The contractor's manner of performance similarly will be adversely affected due primarily to acceleration, lost productivity and efficiency, and delay. This will result in increased construction costs which the contractor must absorb. See **§ 7.6**.

Often the contractor's only hope to survive is a successful lawsuit that allocates the risk of a changed condition to the owner, as discussed in **Chapters 8** and **9**.

This chapter serves as an overview of the evaluation, preparation, and prosecution of differing site conditions claims. The principles discussed in this chapter apply whether the contract contains a changed conditions clause or not.

§ 7.2 Claim Preservation

Unfortunately, for some contractors any claim they may have had as a result of encountering a differing site condition is lost before any initial investigation can be performed. Be it a public or private construction project, most construction documents contain a notice requirement. The contractor's failure to adhere to this condition precedent may result in a forfeiture of its claim. Attorneys must therefore instruct their clients to carefully comply with the contractual notice provisions as a matter of their usual practice.

A contractor would be wise to review the contract documents with its attorneys after the awarding of the contract but before actual construction commences. Some would advocate a review even before the bid is submitted so the contingency of risk could be properly evaluated. Such a review would allow the creation of a simple checklist that would indicate all contractual notice requirements. Aided by this checklist, the contractor, if it encountered what it perceived as a differing site condition or other changes (such as extra work), would at least preserve any claim it might have. Such a procedure allows the contractor and its attorney to more fully investigate the claim and to develop a theory upon which the contractor may be compensated for the added expenses resulting from the unforeseen changed condition.

The importance of strict compliance with contractual notice provisions cannot be overemphasized. Each contract is different with respect to how notice is to be given and to whom the notice must be addressed, as well as the time in which notice must be given after encountering the differing site condition. For example, the differing site conditions clause found in federal construction contracts provides:

> (a) The Contractor shall promptly, and before the conditions are disturbed, give a written notice to the Contracting Officer of (1) subsurface or latent physical conditions at the site which differ materially from those indicated in this contract, or (2) unknown physical conditions at the site, of an unusual nature, which differ materially from those ordinarily encountered and generally recognized as inhering in work of the character provided for in the contract.
>
> (b) The Contracting Officer shall investigate the site conditions promptly after receiving the notice. If the conditions do materially so differ and cause an increase or decrease in the Contractor's cost of, or the time required for, performing any part of the work under this contract, whether or not changed as a result of the conditions, an equitable adjustment shall be made under this clause and the contract modified in writing accordingly.[1]

It is essential to carefully review the contract documents and meticulously draft the checklist accordingly. Such a procedure, although tedious, will not only insure the preservation of claims, but it will also save the contractor the time and expense of trying to justify its failure to give proper notice and/or to prove the owner had actual knowledge of the occurrence even without proper notice.

To prevent the forfeiture of a claim simply because the contractor has failed to strictly comply with the notice provision, some courts have developed exceptions. The most often employed are (1) actual knowledge and (2) waiver or estoppel.[2] Under the first exception, the court balances the actual jobsite activity against the need to comply with the formal notice requirement. The burden is on the contractor to establish that the owner's on-site representative had actual notice of the changed condition, thereby satisfying the true intent behind the formal notice requirement.[3] A waiver or estoppel may be established by demonstrating a course of dealing between the owner and the contractor. In order to succeed and avoid the forfeiture, the contractor must demonstrate that in the past, the owner had not required compliance with the notice provision and that the contractor relied upon the past course of dealing. To successfully establish a waiver or estoppel argument, the contractor must offer examples of when the owner, notwithstanding the contractor's failure to give proper notice, nonetheless paid the contractor for the additional cost as a result of a differing site condition.[4]

---

[1] 4A McBride & Wachtel, Government Contracts § 29.20 (rev. ed. 1991).

[2] *Id.* § 29.50.

[3] Metropolitan Sewerage Comm'n v. R.W. Constr., Inc., 241 N.W.2d 371 (Wis. 1976).

[4] Lawrence-Lynch Corp. v. Department of Envtl. Management, 458 N.E.2d 332 (Mass. 1983).

These exceptions, however, succeed or fail with the facts of any particular situation. There are many cases that uphold the sanctity of the contract and will not deviate from strict compliance with the notice requirements, notwithstanding actual notice or waiver/estoppel.[5]

The prudent contractor will therefore not gamble with the possibility of establishing an exception, but will insure that it strictly complies with all the notice provisions of its contract.

Once the contractor has preserved its claim for a changed condition by complying with the notice provision, the next step is the investigation and development of the facts that support the claim. Once again the starting point is the contract documents. The contractor and/or its attorneys must carefully review and analyze the contract documents.

To recover for a Type I differing site condition, whether under a changed conditions clause or under one of the common law remedies, the contractor must make an initial determination that the actual site conditions materially varied from the conditions represented by the contract documents. It is therefore fundamental that any investigation begin with a determination as to what type of site conditions were represented by the contract documents.

Similarly, to successfully prosecute a Type II changed condition, a review of the contract documents is essential to establish that the site conditions encountered were unknown. This can be established upon a showing that the actual site conditions encountered could not have been reasonably anticipated from the contract documents or a reasonable pre-bid site investigation.

§ 7.3 —Using Independent Consultant

Although the contractor's own conclusions concerning the anticipated site conditions represented by the contract documents are probative on the issue of the existence of a changed condition, the better procedure is to retain an independent consultant. More often than not, the contractor's conclusions are somewhat biased because of the contractor's personal involvement with the project. Personality conflicts that might have arisen between the contractor and the owner's on-site representative may skew the contractor's perception of what is depicted on the contract documents. In addition, should the matter go to trial, the credibility of the contractor's testimony is subject to impeachment based upon the contractor's financial interest in the outcome. The finder of fact may not give the contractor's testimony regarding the representations made by the contract documents the weight necessary to sustain a cause of action. It is foolish to risk losing a legitimate cause of action merely to save the few thousand dollars necessary to engage a competent independent expert.

---

[5] W.&O. Constr. Co. v. City of Smithville, 557 S.W.2d 920 (Tenn. 1977).

It is also good practice for the contractor's attorney rather than the contractor to retain the expert. This allows the expert to freely provide the attorney with the expert's work product without waiving the attorney work product privilege.

The consultant hired should be well respected in the industry. The more well known the expert, the more weight the trier of fact is likely to attach to that individual's testimony and the less likely adverse counsel will be able to attack the witness's credibility.

Document review. Once an expert is selected, counsel must provide the consultant with all the available information regarding the site condition. Here it is wise to err on the side of providing too much information rather than not enough. Thus, the expert must be privy to all the information that was made available at the time of the bid, whether the contractor used this information or not. In addition, counsel should check the history of the site. Was there any prior construction at the site? What about in the vicinity? The attorney should talk to the contractors on those jobs. What type of site condition did they encounter? Was what they encountered substantially similar to the conditions the client encountered? Counsel should ask to see their contract documents. Do they contain similar representations? The expert must have all the information that was at the owner's disposal. If the expert concludes that the contract documents misrepresented the actual site condition, and that the owner knew or should have known of the actual conditions, then the contractor will be in a better position to negotiate a favorable settlement. Such a conclusion would also justify a common law cause of action for fraud and/or misrepresentation. See **Chapter 6**.

Adverse counsel undoubtedly will seek to depose the plaintiff's expert, in preparation for which counsel will demand production of any of the expert's reports or memoranda. Therefore, it is a good practice not to retain any draft copies. Early drafts that contain inconsistencies with the final report are a great source of impeachment materials. To prevent this possibility, clients should institute a corporate policy to dispose of drafts once a report is finalized.

Site investigation. However, one should not limit the expert's investigation to document review. A contractor will be denied recovery for a differing site condition if a pre-bid site investigation would have discovered the actual site conditions, notwithstanding the representation contained in the contract documents. Thus, to the extent it is not disturbed by the construction that has taken place, the expert should make a careful and detailed site investigation in an attempt to determine whether the contractor could have detected the differing site conditions during a pre-bid site investigation. Often the attorney can do nothing to make up for the contractor's failure to do a proper pre-bid site investigation. It is therefore important for the attorney to instruct clients to do a site inspection for every job they intend to bid. These pre-bid investigations should be carefully documented with written reports and pictures. The use of pre-bid pictures is often the only means by which an expert can determine, after the fact, what a proper pre-bid site investigation would have revealed. This process may be time consuming and expensive; however, it is essential to preserving claims.

If a thorough pre-bid investigation was conducted and documented, it can be compared to the expert's own investigation. The expert can effectively use the pre-bid investigation documentation to demonstrate that the condition either was not anticipated or was of an unusual nature.

Once it is determined that there was a changed condition (a physical condition not anticipated by the contract documents (Type I) or an unknown physical condition of an unusual nature (Type II)), the next step is to determine whether that change is actionable, and draft the complaint.

§ 7.4 Preliminary Analysis

In today's climate, it is more important than ever to conduct a preliminary analysis of any potential claim before commencing a lawsuit. This is true for both the owner and the contractor.

For the contractor, a preliminary analysis will safeguard against the possible imposition of sanctions for filing a frivolous lawsuit. Similarly, a preliminary analysis will help the owner make an informed decision regarding a possible settlement.

The first step in any analysis is to determine the theory upon which relief will be sought. The element of each cause of action varies from theory to theory and jurisdiction to jurisdiction; thus, counsel must determine what the basis of the claim will be. An often overlooked element in any cause of action is damages. Therefore, any analysis must be twofold, including (1) proof of entitlement (liability); and (2) proof of loss (damages). The contractor must establish both prongs to succeed at trial.

Before drafting a complaint, the attorney must first know under what theory of law the client is seeking to recover. A prudent attorney will take advantage of the liberal pleading requirements that allow for the pleading of alternate causes of action.

Counsel should begin by simply listing the elements necessary to the cause of action. If the contract contains a differing site conditions clause, the list of elements for a Type I changed condition would include:

1. Certain subsurface or site conditions were indicated on the contract documents
2. The contractor reasonably relied on the representation made by the contract documents
3. The subsurface conditions encountered materially varied from those indicated
4. The contractor suffered damages (increased costs) as a result of the change
5. The contractor gave proper notice under the contract.

The Type II changed condition does not require a comparison of what was represented by the contract documents versus what was actually encountered. To establish a Type II changed condition, the contractor must demonstrate that the condition encountered was an "unknown physical condition . . . at the site, of an unusual nature which differ[ed] materially from those ordinarily encountered and generally recognized as inhering in work of the character provided for in the contract."[6]

It is important to remember that this condition be both unknown and unusual. The first prong of this test is satisfied if the contractor can demonstrate that the condition could not have been reasonably anticipated from the contract documents or from a reasonable site investigation. To establish the second prong, the condition must be unexpected. This does not require a showing of a freak of nature. All that is required is proof that the actual site condition could not have reasonably been anticipated.

Accordingly, the list of elements for a Type II changed condition would be:

1. The contractor encountered a physical condition
2. The condition was unknown to the contractor at the time of bidding
3. The condition could not have been anticipated from the contract documents
4. The condition could not have been discovered by a reasonable pre-bid site investigation
5. The condition was such that an experienced contractor working under similar circumstances would not normally expect to encounter such a condition.

If the contractor is seeking recovery under one of the common law causes of action, such as fraud, misrepresentation, or breach of warranty, the elements will vary slightly. It is important to check the elements of the cause of action in the relevant jurisdiction and plead accordingly.

The attorney should discuss each element with the client, reviewing documentation to support each allegation. This exercise is also helpful to determine the relative strengths and weaknesses of the case. It will indicate early on what problems counsel faces and thus provide the opportunity to take corrective measures.

If the preliminary analysis indicates that the anticipated claim is significant (several hundred thousand dollars), the authors strongly recommend the use of an audit of the alleged damages. The audit should show the backup for each of the dollars being claimed to the extent that such proof is possible. An audit performed by an independent consulting firm can prove to be an invaluable tool to achieve a favorable settlement. Often the increased settlement or judgment will more than make up for the cost of the audit.

---

[6] 4A McBride & Wachtel, Government Contracts § 29.20 (rev. ed. 1991).

The audit also adds instant credibility to the contractor's claim. In addition, the audit provides the attorney with a clear and concise picture of the contractor's actual loss, without any assessment of liability.

§ 7.5 Factual Investigation

As with all construction cases, the earlier a factual investigation begins, the better prepared the case will be for trial. The primary evidence in such cases is the testimony of the personnel who worked on the project. Do not wait until the matter is on the court's calendar to interview potential witnesses. Memories fade with time, so it is essential to interview the potential witnesses as soon as possible after encountering the changed condition.

It is also important to establish a point of contact. Due to the large volume of documents produced in the average construction litigation, one key is to minimize duplication. Thus, the contact point between the attorney and the client is the person through whom all communication with the contractor will pass.

During the course of the litigation, it is also wise for the attorney to hold regularly scheduled meetings with the client. This will insure the client is fully advised of the progress of the litigation and will take advantage of the client's expertise and knowledge of the project. A helpful by-product of these meetings is client education on the reality of the law. For example, such discussions can establish company policy for future jobs regarding compliance with notice requirements and pre-bid site investigation. Such policy can minimize the recurrence of mistakes and help insure the success of future claims.

Investigation. Initial interviews will be with home office personnel. The attorney should determine what, if any, on-site involvement they had, making a list of those individuals who were field personnel along with their titles and to whom they reported. This list should not be limited to those individuals who worked for the client. This list of personnel should also include the names of the companies for whom they worked. This aids in preparing the pleadings as well as in keeping track of discovery demands and understanding the significance of communications among the on-site personnel.

When interviewing the client's field personnel, it is good practice to do so without their boss being present. If the boss is present, employees tend to say what they think the boss wants to hear, which may not necessarily be the truth. If the witness does not mind, tape-recording these interviews will insure an accurate record. If taping is not possible, the attorney should take careful notes.

When questioning jobsite personnel, one should determine their work experience. It is best to use very open questions, such as: Were there any problems? Counsel should let the witness talk. Initially, one should not ask leading questions, but should go through the whole job before asking specifics. This procedure usually helps witnesses refresh their memories. Counsel should remember to ask the witnesses whether they have any documents relevant to the job. Often, field personnel will take photographs while working or maybe maintain a personal diary or journal. These may be useful to the case.

Records. After the preliminary interviews are completed, the next step is to analyze and digest the client's records. Although there are exceptional computer software programs on the market that will aid in the organization of claim files, such litigation support systems are not a substitution for preparation.

If a litigation support system will be used, it is important to reach a consensus regarding who will input the information and what information will be stored. This task is far too important to initially put it in the hands of a clerk or part-time help. The attorney will rely on this information to prove the case. Only someone closely involved with the project has the knowledge to do this job accurately. In addition, the contractor should not do it without the attorney's supervision. The contractor may know the job, but the contractor does not know the law. The attorney must make sure that the facts needed to prove the case are input properly.

The size of the claim and the number of documents involved will determine whether or not to computerize. Once the decision is made to computerize, it is often more cost efficient for the attorney to simply input into her own computer an index or digest of the client's correspondence that is produced jointly. A well-organized file, together with a detailed index, is usually all that is needed. The index should list the relevant correspondence by date, to whom it was written (along with who was copied), by whom it was written, and a detailed reference of subject matter. The reference of subject matter is probably the most important part of the inputting and should be done by both the attorney and the contractor.

To determine which documents are to be indexed, whether manually or aided by a software package, someone should determine which documents are relevant to the claim and how numerous they are. The best procedure is to have both attorney and contractor perform this initial review. This will act as a check and minimize the risk of something falling through the cracks. At this point in the case, it is better to include too many documents rather than not enough. As the case progresses, unnecessary documents can be deleted from the index.

An additional benefit of having the attorney participate in the creation of these digests is that they are protected from disclosure based upon the attorney work product exclusion.

Some practitioners advise numbering the client's documents for inventory purposes. The authors disagree and strongly recommend against it. One can easily retrieve a document based upon its date. The numbering system not only does not add anything, but it acts as a red flag for one's adversary. Remember that eventually these files must be produced for adverse counsel's inspection. If the files are stamped with a number, counsel could easily determine that documents were pulled, which may result in costly motions even if there was a legitimate basis for removing the documents. The cost of litigation is high enough. One should seek to avoid needless motion practice.

Discovery. Once the review and analysis of the client's files is completed, the next step is to inspect the files of one's adversary and any third parties. Remember that counsel is allowed to subpoena documents from nonparties, such as consultants, inspectors, and suppliers. The factual investigation should not stop at the defendant's files.

Armed with the list of potential witnesses prepared during the initial interviews, the attorney should sit down with the client and discuss what documents are available and from whom. It is important to be as specific as possible to avoid being subject to a motion for a protective order on the grounds that you are overreaching. Such a motion represents added cost and can delay the prosecution of the case. The attorney must not give the impression she is on a fishing expedition, thus weakening the plaintiff's negotiation position.

Whether the attorney has retained an independent consultant or is using the client's expertise, those resources are a valuable means of developing and drafting discovery demands. Their focus should be primarily on insuring that demands are properly formulated to obtain the documentation needed to prove the claim.

Another valuable source of information is the public owner. As-built drawings, permit applications, and other public filings may be helpful to establish whether the contractor's reliance upon the representation contained in the contract documents was reasonable or whether there was some information that was withheld. Assuming the municipality is not the owner of the project, the attorney must check the civil practice rules in the particular locality regarding the procedure to be followed to obtain the documents. Some jurisdictions require the issuance of a court order before a subpoena can be served on a municipality. If the municipality is a party to the action, a simple notice to produce will be sufficient.

However, if the municipality is a nonparty, one may be able to save time and money by contacting its attorney. Often, because the municipality has no direct interest in the lawsuit between the owner and the contractor, it will stipulate to the issuance of the subpoena and agree to produce the requested documents.

Another option on a public contract is to make a demand via the Freedom of Information Act.[7] The attorney should check the statute to make sure the information wanted is accessible.

While demands for production of documents are pending, counsel should develop a procedure for digesting and filing these documents. The authors strongly recommend maintaining a control file to insure that all the documents are in one place and that one can identify who produced what documents.

---

[7] 5 U.S.C. § 552 (1988).

At this point, the use of a numbering system is helpful. One can establish a simple code to identify from whom the documents were received. By stamping all the documents in such a way, it is easy to identify the source of any given document.

Depositions. Once all the documents necessary to establish the claim are obtained, the next step is to determine if a deposition would be necessary and, if so, of whom. In this regard, the client's input is important. The client has worked on the project and will know who would be sympathetic to the cause. Keep in mind, however, that the purpose of depositions is not only to support the claim, but also to learn the facts on which one's adversary will rely.

In preparation for the depositions, the attorney should review the pleadings. Assuming one has pled a valid cause of action, the burden of proof is limited to those allegations in the complaint that the defendant denied. The attorney should not waste time and resources on allegations the defendant has already admitted.

The attorney should prepare for a deposition with the same intensity that counsel prepares for trial. Witnesses have been known to die or disappear, in which case deposition transcript is critical. Therefore, the attorney must treat the deposition as the witness's testimony, making sure to cover everything needed to establish the claim or defeat the defense.

Today many jurisdictions allow attorneys to videotape a deposition. Although this is more costly, if the witness is ill or a risk to disappear, it is recommended. Often these cases come down to a swearing contest, with the trier of fact ultimately deciding who to believe based partially on the witness's demeanor. If there is a key witness whose testimony is absolutely essential, counsel must determine if that person's testimony would carry the same persuasive effect if it was read into the record as opposed to being viewed. The use of video depositions is a decision the client must make, but the attorney should make sure the client is aware of all possible options.

Once the depositions are completed, the attorney should summarize and digest the transcripts for ease of reference. Counsel should select the individual who will digest the transcripts wisely. The person must be well versed with the underlying facts of the dispute as well as the legal issues to be decided. The attorney should review the summaries and transcripts with an eye toward confirming admissions that may have been made.

Review. At this point, the attorney should review the pleadings once again to see which allegations are left to be proven. Counsel should take a clear copy of the complaint and mark all the allegations one's adversary has admitted, noting how it was admitted (such as in the answer or in a deposition). Such an indication will help in drafting a summary judgment motion if one can establish the case as a matter of law. If not, counsel might serve upon the adversary a request for admission. This is an underutilized but powerful device to limit the issues to be tried. Not only may one get the adversary to make further admissions, thereby lessening one's burden of proof, but the request for admission is also a great way of solving any evidentiary problems related to certain documentation. One can, and should, use the request for admission to have one's adversary authenticate documents and to avoid any foreseeable hearsay problems, such as admission of business records.

The proper use of a request for admission will avoid the need to examine witnesses on tedious foundation issues, thereby shortening the length of the trial.

Trial preparation. Every practitioner prepares for trial in a different way; however, there are certain concerns that run through all preparation.

The attorney must meet with the witness. Here, as in all cases, counsel must balance the need for thoroughness against the possibility of the witness becoming overly prepared. The witness must not come across as a parrot who simply repeats what the attorney had told her to say. This must be avoided at all costs. Not only is it possibly unethical, but it is also not effective. The trier of fact will sense what is being done and will immediately discount that witness's testimony.

It is wise to meet with each witness individually. It is also a good idea not to have the client present. The witness will be more comfortable and direct in private.

The attorney should get the witness to talk, rather than ask leading questions. Such questions will not be permitted at trial, so counsel should begin now by asking very broad questions to get the witness to tell her story. The witness will come across as more credible than if she simply answers direct questions. Once the witness has completed her story, counsel can go back and ask specific questions to fill in the gaps.

The attorney should spend a few minutes explaining to the witness the trial procedure. The witness should know where the court is located and in which courtroom, as well as who will be present while she is testifying. This may be the witness's first contact with the litigation process, and she will undoubtedly be nervous. The attorney should do whatever is possible to relieve the witness's anxiety.

The witness should not leave before giving counsel her telephone numbers (home, work, car, beeper). The attorney must be able to reach the witness at all times. The attorney should discuss with the witness her vacation plans, lest one be on trial and learn that a key witness is out of town.

Once witnesses are lined up and have been briefed, the attorney should discuss with the client the order in which they should testify. If possible, it is convenient to take the witnesses in chronological order through the job in the sequence it was constructed. If, however, the project does not lend itself to such a procedure, the attorney should keep in mind when positioning witnesses that most people remember the first thing and the last thing they hear. Therefore, the first and last witnesses should be the best.

The next step is to marshal the exhibits, going through the testimony of each witness and putting the exhibits to be introduced in order. There should be enough copies for each attorney and a copy for the court. Counsel should examine each document carefully to anticipate any objection to its admissibility and should have an outline to insure laying the proper foundation.[8]

§ 7.6 Proving Damages

Recovery partially depends upon the attorney's ability to show how the changed condition impacted on the contractor's work. The contractor should thoroughly document added costs, including what new equipment was needed, how many additional workers were used, and how much added time was involved. Whatever form these records take, such as time sheets or foreman's reports, they should be maintained on a daily basis. This not only insures their accuracy, but it may also be mandatory under the contract. In some construction contracts, the contractor is required to submit, on a daily basis, force account records for any disputed work.

The attorney should establish that the contractor's bid was reasonable. To do this, it is often useful to compare the bid to that of other bidders or to the owner's pre-bid estimate. Although such comparisons are probative, they are not controlling. The ultimate question is whether the contractor's bid was reasonable based upon the representation contained in the contract. That is, what would a reasonable contractor have bid in the same situation? Consideration should be given to getting a competitor or an expert to testify as to the reasonableness of the contractor's bid.

Damages caused by a differing site condition are often difficult to prove. Rarely, if ever, will payment of direct costs alone fully compensate the contractor. Owners often fail to appreciate the ripple effect such a change has on other contract operations. Often a differing site condition will cause not only a period of delay, but will also create a period of constructive acceleration. The contractor, if not granted a time extension, will be forced to accelerate its operation so as to avoid the assessment of liquidated damages. There is also an effect on scheduling and coordination of the various trades that results from the different site conditions that can be categorized as lost productivity.

The basic principle in computing damages is that the contractor should be fully reimbursed for all costs incurred as a result of the changed conditions. These damages should include (1) direct costs (labor and equipment), (2) impact costs, and (3) time extension. The time extension is important because it

[8] The authors strongly recommend *Evidentiary Foundations* by Edward J. Imwinkelried (1980). It will solve almost any foundation problem you think of and a few you did not even know existed.

releases the contractor from its obligation to complete the project by a specific time. More important, however, it will serve as the basis for the return of any liquidated damages that may have been assessed.

A commonly used formula for computing lost productivity is the use of the measured mile. That is, compare the contractor's productivity during a period of time in which it was not adversely affected by the differing site conditions to the productivity when the contractor was hampered by the differing site conditions. This method of calculation is widely accepted as objective.

To use this measured-mile approach successfully, a fundamental understanding of the work operation is essential. The work performed in both periods must be, if not identical, substantially similar. Next, one must be able to minimize the effects other factors had on the contractor's productivity. The measured-mile comparison may not be valid if one compares the productivity of excavating a differing site condition during the month of February with excavating what was anticipated during the month of April. There may be too many dissimilarities between the two periods to enable a successful comparison. The ground is harder in the winter, and the natural productivity is decreased in the cold weather. The measured mile must reasonably recreate the factors that were present during the encountering of the changed condition.

Once one has selected the periods of time to compare, the next step is to compute the productivity factor. Productivity can be measured in many ways, such as production rate per day, per week, or per hour. The authors suggest using the unit of measurement utilized in the contractor's bid. Such a comparison can also serve to demonstrate the reasonableness of the bid.

One problem that often arises is the inability to isolate a measured mile. Often the differing site condition is so pervasive that there was no period of time during which the contractor was unaffected by the changed condition. In such a case, one can still use the measured mile approach; however, by the very nature of the change, damage calculations are subject to attack based upon other variables that contributed to the loss of productivity.

In this situation, counsel should seek the help of an expert, be it an engineer, consultant, another contractor, or other individuals knowledgeable in the construction industry who can testify what, if any, effect those other variables had on the contractor's productivity.

One could also compare the contractor's bid estimate to the actual cost. This, however, requires a finding that the bid was reasonable, not often an easy fact to establish. It places the entire bid and backup into controversy. Adverse counsel will scrutinize all pre-bid determinations in an attempt to establish that the bid's productivity estimate is not a reasonable measurement of what should have been anticipated.

An alternate to going back to the bid is to pick a period of time in which the contractor was least impacted by the differing site conditions and compare it to the affected portion of the work.

Another method of computing damages for lost productivity is to show actual increased costs. This, however, cannot be done in hindsight. It requires the contractor to segregate the costs of affected work once it encounters the differing site condition. By segregating these costs, they can be easily verified through time cards and invoices. The problem with this method is twofold. First, contractors often cannot make such a segregation, and to do so is costly and time consuming. Second, the method does not fully compensate the contractor for its total cost. Hidden costs (such as certain aspects of home office overhead), which are not calculable, are lost. This method does not take into account the ripple effect such extra work has on contract work.

Often the decision regarding which method of calculation of damage will be followed is made before the attorney is retained. The attorney can do nothing about the client's recordkeeping procedure except to make the best of it.

CHAPTER 8

RECOVERY UNDER TYPE I AND TYPE II CLAIMS

Edward A. Hannan

Edward A. Hannan is a general partner in Godfrey, Braun & Hayes of Milwaukee, Wisconsin. His practice emphasizes the construction industry with primary emphasis upon the representation of design professionals, construction product manufacturers, and insurers. He received his law degree from the University of Wisconsin-Madison. Mr. Hannan is a member of the Wisconsin bar, the American Bar Association, American Arbitration Association, Construction Specification Institute, and the State Bar Association of Wisconsin.

§ 8.1 Introduction

Differing site conditions clauses provide for making cost adjustments between contractors and owners in response to increases or decreases in anticipated performance cost. Generally, it is the contractor who seeks adjustment under the clause. The condition most often complained of is one that differs materially from those stated in the contract. These claims are referred to as *Type I claims.* Change conditions may also consist of unknown conditions of an unusual nature, not ordinarily encountered in the kind of work provided for in the contract. Claims that fall within this area are categorized as *Type II claims.*

The vast majority of precedent construing and applying differing site conditions clauses is based on interpretations of the standard Federal Acquisition Regulation's (FAR) differing site conditions clause, which is required for all federal projects.[1] Therefore, most case law in this area involves public works contracts. The most prolific source of this law is found in the various boards of contract appeals decisions.

Decisions of the contract appeals boards resolve disputes between the federal government procuring agencies and private contractors. The jurisdiction of the contract appeals boards arises under the Contract Disputes Act of 1978, which is implemented by the disputes clause in government contracts.[2]

When dealing with the United States government, these administrative avenues of redress must be exhausted before a claim can be brought before a legislative court.[3] The Wunderlich Act governs judicial review of these administrative decisions.[4] Under the Wunderlich Act, administrative conclusions of law are not binding on the reviewing court, but administrative findings of fact are binding, unless fraudulent, capricious, so grossly erroneous as necessary to imply bad faith, or not supported by substantial evidence.[5]

In regard to differing site conditions clauses, the question of whether a contractor encountered a changed condition is a question of fact, but an issue regarding an equitable adjustment under the changed conditions provisions is a question of law when it involves contract interpretation.[6]

Review of administrative decisions, such as those rendered by a board of contract appeals, may be obtained through the Court of Claims.[7] The jurisdiction of the Court of Claims extends throughout the United States.[8] Therefore, it does not matter where the claim originated or where the parties reside, as long as the claim is within the subject matter jurisdiction of the court.[9] The Court of Claims has jurisdiction upon a claim against the United States based upon a contract with the United States.[10] Generally, the Court of Claims does not have

---

[1] F.A.R. 52.236-2 (1984); F.A.R. Form 23A.

[2] 41 U.S.C. § 601 (1982).

[3] United States v. Joseph A. Holpuch Co., 328 U.S. 234, 240 (1946).

[4] 41 U.S.C. §§ 321–322 (1964).

[5] 64 Am. Jur. 2d *Public Works and Contracts* § 121 (1972).

[6] *Id.*

[7] 28 U.S.C. § 171(a) (1982).

[8] 32B Am. Jur. 2d *Federal Practice and Procedure* § 2555 (1982).

[9] *Id.*

[10] 28 U.S.C. § 171(a) (1982).

jurisdiction over a claim by a subcontractor because there is no privity of contract between the subcontractor and the United States.[11]

Federal precedent respecting differing site conditions area has also been developed by the United States Court of Appeals for the Federal Circuit, which reviews decisions of the United States Court of Claims.

The 1987 edition of the American Institute of Architects (AIA) Document A201 basically adopts the provisions of the FAR differing site conditions clause.[12] Accordingly, the vast body of federal precedent on differing site conditions will now be available to guide state courts in interpreting and applying the current AIA clause.

However, counsel should exercise caution when using case law interpreting the FAR differing site conditions clause in a scenario not including the federal government as a party. This is especially true when one uses decisions of a board of contract appeals to support an argument. A decision of a board of contract appeals may rest substantially upon government procurement regulations and stand apart from many of the cases developed under commercial contracts between private parties.[13] Practically speaking, however, decisions of the boards of contract appeals interpreting and applying differing site conditions clauses may prove to be persuasive authority to the extent that a pattern of construction exists.

The burden of proof for both a Type I and Type II claim is on the contractor.[14] The decisional standard for both categories is preponderance of the evidence.[15]

§ 8.2 Type I Claims

A Type I differing site condition exists when "subsurface or latent physical conditions at the site differ materially from those indicated in the contract."[16] To succeed in a Type I differing site conditions claim, the contractor must

---

[11] 32B Am. Jur. 2d *Federal Practice and Procedure* § 2561 (1982).

[12] Hannan, *What's Different about Differing Site Condition Claims?* 57 Def. Couns. J. 304, 306 (1990).

[13] *How to Use This Reporter,* 92-1 B.C.A. (CCH) 5 (1990).

[14] Hannan, *What's Different about Differing Site Condition Claims?* 57 Def. Couns. J. 306 (1990); Medsger, *Category II Differing Site Conditions in Construction Contracts,* Army Law., June 1988, at 10, 11.

[15] Medsger, *Category II Differing Site Conditions in Construction Contracts,* Army Law., June 1988, at 11.

[16] J.M.J. Invs., AGBCA Nos. 87-259-1, 87-341-1, 91-3 B.C.A. (CCH) ¶ 24,072, at 120,510 (1991).

demonstrate "that the conditions 'indicated' in the contract documents differ materially from those it encounters during performance."[17]

These claims typically arise from a subsurface physical site condition, such as groundwater, rock, or unsuitable soils,[18] but they are not limited to these and may involve latent conditions of any type.[19] Additionally, a differing site condition claim could arise from quantities of materials or substances contrary to what was indicated in the contract documents.[20]

To help clarify what the contractor must show to be successful under a Type I claim, boards of contract appeals have articulated various tests. Essentially, to prevail a contractor must show:

1. A subsurface or latent physical condition
2. Located at the site
3. That is materially different from those indicated in the contract documents.[21]

Eventually, some boards of contract appeals and courts elaborated upon these elements to more fully explain the analysis. The following is an example of one such test:

1. Actual conditions at the site differing materially from those indicated in the contract
2. Which conditions could not have been ascertained by a site investigation and a pre-bid review of the contract
3. The contractor's reliance on its interpretation of the contract
4. Damage.[22]

More recently, a board of contract appeals has employed a six-element test when reviewing Type I claims. This test, although somewhat more refined, does not substantively add any elements not previously mentioned. These elements are:

1. Contract documents affirmatively indicate the subsurface or latent conditions forming the basis for the claim

---

[17] Clark v. United States, 5 Cl. Ct. 447, 451 (1984), *aff'd,* 770 F.2d 180 (Fed. Cir. 1985); P.J. Maffei Bldg. Wrecking Corp. v. United States, 732 F.2d 913, 916 (Fed. Cir. 1984).

[18] Fattore Co. v. Metropolitan Sewage Comm'n, 454 F.2d 537 (7th Cir. 1971).

[19] Skip Kirchdorfer, Inc., ASBCA No. 23443, 79-2 B.C.A. (CCH) ¶ 14,092 (1979).

[20] J.M.J. Invs., AGBCA Nos. 87-259-1, 87-341-1, 91-3 B.C.A. (CCH) ¶ 24,072 (1991).

[21] Foster Constr. C.A. v. United States, 435 F.2d 873 (Ct. Cl. 1970).

[22] Brechan Enters., Inc. v. United States, 12 Cl. Ct. 545 (1987); *see also* EZ Constr. Co., ASBCA No. 31506, 87-1 B.C.A. (CCH) ¶ 19,679 (1987).

2. The contractor must have acted as a reasonable and prudent contractor in interpreting the contract documents
3. The contractor must have relied on the indications of the subsurface or latent conditions in the contract
4. The subsurface or latent conditions actually encountered at the site must materially differ from the contract indications
5. The actual conditions encountered must have been reasonably unforeseeable
6. The costs claimed must be solely attributable to materially different subsurface conditions.[23]

Each of these six elements is individually studied in §§ **8.3** through **8.7**. General rules and specific case examples are referenced when appropriate.

§ 8.3 —Affirmative Indication of Differing Site Condition

"To prevail on a Type I differing site condition, there must be reasonably plain or positive indications in the contract documents that conditions were otherwise than that actually encountered."[24] "The determination of what the Contract documents 'indicate' with reference to a subsurface or latent physical condition is a question of contract interpretation."[25]

The contractor is required to review all documents including noncontract documents to determine what the contract indications consist of. "Contract 'indications' generally include information both in the solicitation itself and in documents to which bidders are directed by the solicitation."[26] *Contract documents* include plans, details, specifications and general and specific conditions, soil boring logs or soil profiles, fitting schedules, estimates in bid documents, aerial photographs, and any interpretive notes or commentary in or accompanying the contract or bid documents.[27]

---

[23] Simpson Constr. Co., VABCA No. 3176, 91-1 B.C.A. (CCH) ¶ 23,630, at 118,390 (1990); Shumate Constructors, Inc., VABCA No. 2772, 90-3 B.C.A. (CCH) ¶ 22,946 (1990); Weeks Dredging & Contracting, Inc. v. United States, 13 Cl. Ct. 193 (1987), *aff'd,* 861 F.2d 728 (Fed. Cir. 1988).

[24] J.M.J. Invs., AGBCA Nos. 87-259-1, 87-341-1, 91-3 B.C.A. (CCH) ¶ 24,072 at 120,510 (1991).

[25] Shumate Constructors, Inc., VABCA No. 2772, 90-3 B.C.A. (CCH) ¶ 22,946 at 115,193 (1990).

[26] Ashbach Constr. Co., PSBCA No. 2718, 91-2 B.C.A. (CCH) ¶ 23,787 at 119,134 (1991).

[27] Hannan, *What's Different about Differing Site Condition Claims?* 57 Def. Couns. J. 304, 308 (1990).

If two or more of these documents create an ambiguity, there is a patent defect, and the contractor has a duty to inquire.[28] A simple example of such a scenario included a case in which the general requirements clause of a contract directed the contractor to include the cost of removing a utility pole. This requirement was also indicated in the electrical site plan. However, a survey plan stated all overhead utilities had been removed under a separate contract. After denying the differing site conditions claim, a board of contract appeals held that the survey plan created an ambiguity, but it was patent, requiring further inquiry by the contractor.[29]

An apparent exception to the rule that an ambiguity in two or more documents creates a duty to inquire was established in a case in which a contractor was apparently put on notice of certain adverse rock conditions by core borings. A board of contract appeals held the contractor may still have a Type I claim if the contract contains language that would dissipate by unequivocal contradictions the conditions the core borings indicated.[30] In this case, the contradiction was the required use of a blasting method that would not be successful in the type of rock encountered or apparently indicated in the core borings.[31]

If there is no ambiguity in the contract documents, the contractor has less of a duty to inquire. It is well settled that when Type I differing site conditions are involved, a contractor is entitled to rely on the representations of site conditions in the contract and is under no obligation to conduct its own measurements or surveys.[32] A contractor in such a situation is bound only by what a simple site viewing would reveal.[33]

In some situations, certain documents may not be relied on by the contractor and will not be considered part of the contract indications. These cases typically involve documents with some type of warning or disclaimer.[34] In one such case, a contractor damaged a ceiling by drilling through the first floor of a building. There was an error in the thickness of the floor as shown on the "as-built" drawings provided to the contractor. The contractor's claim for repair costs was denied because the as-built drawings were provided at the contractor's insistence, were not part of the contract, and were provided with a warning the building had undergone many modifications since the drawings were prepared.[35]

---

[28] Ashbach Constr. Co., PSBCA No. 2718, 91-2 B.C.A. (CCH) ¶ 23,787, at 119,135 (1991); S.O.G. of Ark. v. United States, 546 F.2d 367 (Ct. Cl. 1976).

[29] Pinner Constr. Co., PSBCA Nos. 2425, 2470, 2543, 91-3 B.C.A. (CCH) ¶ 24,143, at 120,805 (1991).

[30] Sierra Blanca, Inc., ASBCA Nos. 30943, 32286, 34838, 91-2 B.C.A. (CCH) ¶ 23,990, at 120,080 (1991).

[31] *Id.*

[32] Simpson Constr. Co., VABCA No. 3176, 91-1 B.C.A. (CCH) ¶ 23,630, at 118,391 (1990).

[33] *Id.*

[34] P.J. Maffei Bldg. Wrecking Corp. v. United States, 732 F.2d 913 (Fed. Cir. 1984).

[35] Cocoa Elec. Co., ASBCA No. 33921, 91-1 B.C.A. (CCH) ¶ 23,442, at 117,591 (1990).

The contract indications may be based on either express contractual statements or implicit representations.[36] To prevail on a Type I differing site condition, the contractor need not show an explicit indication in the contract, but it must show that the contract contains some reasonable implication that a certain condition existed.[37] The contract indication can be found if the contract documents provide "sufficient grounds to justify a bidder's expectation of latent conditions materially different from those actually encountered."[38] Silence "does not of itself establish a right to recovery."[39] The absence of any information in the contract documents concerning the alleged differing site condition obviates recovery for a Type I condition.[40]

In determining if a condition is implicitly represented, a board of contract appeals or court may look for design requirements that cannot be met or construction methods that cannot be used because of the conditions encountered. In one case, a board of contract appeals granted recovery for a Type I claim when the building excavation work called for a 95-percent compaction of the subgrade; the contract was determined to implicitly "indicate" more favorable soil conditions because a 95-percent compaction rate was impossible due to the excessive water retention characteristics of the soil.[41] Another case considered the fact that the clearing or grubbing work required in the contract was impossible because the equipment necessary for such work would simply sink into the mucky soil encountered.[42]

Lack of detail in the description of a condition referenced in the contract documents may also be found to imply a certain condition exists. In one such case, the contract identified the presence of a six-inch clay tile pipe. The board of contract appeals held that the lack of a label identifying the pipe as a French drain, which is perforated, amounted to an implicit representation that the pipe shown was a solid-walled pipe.[43]

The second significant requirement of the first element is that the subsurface or latent condition forming the basis of the claim is a *physical condition.* This requirement is extracted directly from the language comprising the differing

---

[36] Foster Constr. C.A. v. United States, 435 F.2d 873 (Ct. Cl. 1970).

[37] Fred Burgos Constr. Co., ASBCA No. 41395, 91-2 B.C.A. (CCH) ¶ 23,706, at 118,675 (1991).

[38] Shumate Constructors, Inc., VABCA No. 2772, 90-3 B.C.A. (CCH) ¶ 22,946, at 115,193 (1990); P.J. Maffei Bldg. & Wrecking Corp. v. United States, 732 F.2d 913, 916 (Fed. Cir. 1984).

[39] Jen-Beck Assocs., VABCA No. 2121, 86-3 B.C.A. (CCH) ¶ 19,056 (1986).

[40] Glenn's Heating, ASBCA No. 25754, 83-1 B.C.A. (CCH) ¶ 16,358, at 81,300 (1983).

[41] R.L. Spencer Constr. Co., ASBCA No. 18450, 75-2 B.C.A. (CCH) ¶ 11,604, at 55,423 (1975) (the board also made note of the fact that the water retention characteristics of the soil were not discernible from boring data).

[42] Kinetic Builders, Inc., ASBCA No. 32627, 88-2 B.C.A. (CCH) ¶ 20,657, at 104,400 (1988).

[43] Caesar Constr., Inc., ASBCA No. 41059, 91-1 B.C.A. (CCH) ¶ 23,639, at 118,417 (1990).

site conditions clause.[44] The physical condition requirement is also necessary for a successful Type II claim.[45] In fact, determining if the alleged differing site condition is a physical condition will more likely be an issue in Type II claims.

A *physical condition* has been interpreted to mean an event or condition occurring in nature rather than an economic, governmental, or political condition.[46] However, this "occurring in nature" language is misleading. The physical condition may also be man-made. One distinction that could be made between a Type I and a Type II physical condition is that, although both may be something found in nature, Type II conditions are more commonly man-made conditions that were not recognized or that the owner did not know existed before commencement of the work.[47]

Generally relief has been denied when expense resulted from weather conditions on the ground that intervening weather conditions do not constitute differing site conditions.[48] Courts have routinely held that unexpected, unanticipated, or unprecedented weather or climate conditions are generally deemed to be acts of God not within the purview of the differing site conditions clause.[49]

In several cases, however, although adhering to the rule that adverse weather in itself is not a differing site condition, a board of contract appeals held that certain conditions resulting from that weather may be compensable under the differing site conditions clause. For example, unusual pressure from incoming tides combined with peculiar structural features of a dock, causing continuous seepage onto the dock, was found to be a differing site condition.[50] In another case, although the board denied the claim, it stated that under some conditions interaction of soil and excessive precipitation might constitute a differing site condition.[51] Storm damage to a construction site did amount to a changed

---

[44] F.A.R. 52.236-2(a)(1) (1984).

[45] *Id.* at 52.236-2(a)(2).

[46] *See* Hallman v. United States, 68 F. Supp. 204 (Ct. Cl. 1946); Robert E. McKee Gen. Constr., Inc. v. United States, ASBCA No. 5521, 60-1 B.C.A. (CCH) ¶ 2,526 (1960) (equipment shortage and labor strikes not within purview of "physical conditions"); Keang Nam Enters., Ltd., ASBCA No. 13747, 69-1 B.C.A. (CCH) ¶ 7705 (1969) (evacuation forced by dangers posed by commencement of Tet offensive during Vietnam War did not qualify as "physical condition" for purposes of differing site condition claim).

[47] Medsger, *Category II Differing Site Conditions in Construction Contracts,* Army Law., June 1988, at 15.

[48] 65 Am. Jur. 2d *Public Works and Contracts* § 183 (1972).

[49] Praxis-Assurance Venture, ASBCA No. 24748, 81-1 B.C.A. (CCH) ¶ 15,028 (1981) (abnormal rainfall); Arundel Corp. v. United States, 103 Ct. Cl. 688, 711–12 (1945) (hurricane); Hardeman-Monier-Hutcherson v. United States, 458 F.2d 1364, 1370–71 (Ct. Cl. 1972) (adverse sea and wind conditions).

[50] Warren Painting Co., ASBCA No. 18456, 74-2 B.C.A. (CCH) ¶ 10,834 (1974).

[51] F.D. Rich Co., ASBCA No. 6704, 1963 B.C.A. (CCH) ¶ 3710 (1963); *see also* Frank W. Miller Constr. Co., ASBCA No. 22347, 78-1 B.C.A. (CCH) ¶ 13,039 (1978) (sand deposited uphill from jobsite by another contractor washed down during heavy rainfall).

condition under a contract clause allowing the contractor to request additional compensation.[52]

"No one can ever know with certainty what will be found during subsurface operations."[53] It is this uncertainty that led to the very creation of differing site conditions clauses, as well as an expansive definition of what a *subsurface condition* would include. Examples of subsurface conditions that may qualify as Type I conditions under appropriate circumstances include the presence of subsurface rock formations,[54] the physical or mechanical properties of subsurface rock,[55] the quantity of rock to be excavated,[56] the presence of artesian water,[57] physical properties of subsoils,[58] mechanical properties of subsoils,[59] and subgrade structures.[60]

Latent conditions need not be subsurface conditions. The phrase encompasses any physical conditions that ordinarily escape the attention of a reasonable and prudent contractor, whether below, at, or above grade, including natural as well as man-made conditions existing at the time of or after issuance of the bid documents.[61] A latent condition is any physical condition reasonably unforeseeable to an ordinarily prudent contractor acting in the same or similar circumstances as the contract at issue.[62] Examples include instability of soils forming an embankment,[63] undisclosed concrete piles,[64] undisclosed ductwork,[65] and the thickness of a concrete floor.[66]

---

[52] New Pueblo Constr., Inc. v. State, 696 P.2d 185 (Ariz. 1985).

[53] Kaiser Indus. Corp. v. United States, 340 F.2d 322, 329 (1965).

[54] J.A. LaPorte, Inc., ENG BCA No. 5254, 88-3 B.C.A. (CCH) ¶ 20,953 (1988).

[55] Richard Goettle, Inc. v. Tennessee Valley Auth., 600 F. Supp. 7 (N.D. Miss. 1984) (claim denied for "weathered rock").

[56] Coastal Marine Constr., AGBCA No. 89-190-3, 90-2 B.C.A. (CCH) ¶ 22,895 (1990); *but cf.* Aoki Corp., ENG BCA No. PCC-62, 91-2 B.C.A. (CCH) ¶ 23,848 (1991) (substantial variation from estimated quantities in unit price contract is not sufficient, in itself, to establish differing site condition).

[57] United Contractors v. United States, 368 F.2d 585 (Ct. Cl. 1966); Nelson Bros. Constr. Co. v. United States, 225 Ct. Cl. 703, 650 F.2d 290 (1980) (live subsurface waters).

[58] W.F. Magann Corp. v. Diamond Mfg. Co., 580 F. Supp. 1299 (D.S.C. 1984) (absence of clay binders in sand).

[59] Paul N. Howard Co. v. Puerto Rico Aqueduct Sewer Auth., 744 F.2d 880 (1st Cir. 1984) (soil instability).

[60] Excavation Constr. Co., ENG BCA No. 3646, 77-1 B.C.A. (CCH) ¶ 12,224 (1976).

[61] Hannan, *What's Different about Differing Site Condition Claims?* 57 Def. Couns. J. 304, 307 (1990).

[62] Mojave Enters. v. United States, 3 Ct. Cl. 353, 357 (1983).

[63] Paul N. Howard Co. v. Puerto Rico Aqueduct Sewer Auth., 744 F.2d 880, 883 (1st Cir. 1984).

[64] D. Federico Co. v. New Bedford Redev. Auth., 723 F.2d 122, 125 (1st Cir. 1983).

[65] Joseph Morton Co., GSBCA No. 4611, 80-2 B.C.A. (CCH) ¶ 14,771 (1980).

[66] J.E. Robertson Co. v. United States, 437 F.2d 1360 (Ct. Cl. 1971).

§ 8.4 —Conduct of Contractor

Contract interpretation is conducted from the aspect of a reasonable and prudent contractor to ascertain whether the physical conditions the contractor relied upon do in fact exist as indicated in the contract documents.[67] A contractor will not be deemed to know that which only an expert could derive from the bid documents.[68]

In a case in which the contract described the excavation negatively as "unclassified . . . regardless of the nature of the material excavated," the Claims Court held that a reasonably prudent contractor would not have assumed that no solid rock would be encountered.[69]

§ 8.5 —Reliance

Reasonable reliance is an essential element of a Type I differing site condition. A contractor claiming a Type I differing site condition will not be successful, even if the site conditions vary from the contract, unless the contractor reasonably relies on the contract's assertions as to the condition.[70] This requirement is typical of most misrepresentation claims. If the contractor did not reasonably rely on the contract indications, it did not suffer damages as a result of them and therefore cannot recover.

§ 8.6 —Materiality

The condition under consideration, most naturally, must be located at the jobsite where the contract is being performed. The analysis of whether the location of the alleged differing site condition is at the site will be the same for both a Type I and a Type II claim. The following analysis is applicable in either situation. The site of the work includes any area "designated" in the contract documents or areas where the contractor is otherwise directed to perform work.[71]

Differing site conditions at off-site locations have been sufficient to justify a claim when the owner has designated a specific quarry or boreal site at issue

---

[67] Shumate Constructors, Inc., VABCA No. 2772, 90-3 B.C.A. (CCH) ¶ 22,946, at 115,194 (1990).

[68] Stock & Grove, Inc. v. United States, 493 F.2d 629 (Ct. Cl. 1974).

[69] Fox v. United States, 7 Cl. Ct. 60, 62 (1984), *aff'd,* 785 F.2d 326 (Fed. Cir. 1985).

[70] Gerald Miller Constr. Co., IBCA No. 2292, 91-2 B.C.A. (CCH) ¶ 23,829 (1991).

[71] Hannan, *What's Different about Differing Site Condition Claims?* 57 Def. Couns. J. 304, 307 (1990).

and made representations as to the suitability of materials at such a site.[72] Therefore, an argument that the area where the differing site condition was encountered was part of the site because the area was essential to the performance of the contract may not be successful.[73] Additionally, government approval of the site may not render the owner liable because approval is typically not part of the owner's contractual responsibilities.[74]

The boards of contract appeals and the various courts have had a difficult time determining if the condition encountered is "materially different" from that indicated in the contract. Determining materiality in Type II claims has been equally difficult when comparing the encountered condition with the known and the usual.[75] Because of this difficulty and the fact-intensive nature of differing site conditions claims, boards of contract appeals have resisted defining tests for ascertaining "material" differences from what was indicated in the contract or what was known and usual. Most decisions generally discuss if a difference exists and then summarily conclude that the differing site condition claim should or should not be allowed, without specifically stating why the court felt that the difference was material or immaterial.

The materiality analysis in Type I claims is somewhat less complex than in Type II claims. In a Type I claim, there is some contract indication to compare to the encountered condition. This indication, although maybe not as specific as a quantity or percentage, is usually more concrete than what is "usual," the standard of comparison in a Type II claim.

For example, in *Simpson Construction Co.,*[76] a drywall contractor encountered walls that varied from 15 feet, 3 inches at the low side to 17 feet, 6 inches at the high side. In analyzing the contractor's Type I claim, the board looked to the contract and found that the height was indicated to be 14 feet, plus or minus. After determining that the "plus or minus" referred to inches, the court concluded the actual site conditions differed materially from those indicated in the contract.[77]

---

[72] Blaze Constr. Co., IBCA No. 2863, 91-3 B.C.A. (CCH) ¶ 24,071, at 120,506 (1991) (sacred Indian religious site encountered by contractor did not amount to Type II claim because condition was on leased area used for construction gravel and sand pit and was not part of jobsite). *Compare, e.g.,* Tobin Quarries, Inc. v. United States, 84 F. Supp. 1021 (Ct. Cl. 1949) (quarry designated in contract documents as approved quarry qualified as "site" for purposes of differing site condition claim) *with* L.G. Everist, Inc., 231 Cl. Ct. 1013 (1982), *cert. denied,* 461 U.S. 957 (1983) (quarry not designated in contract documents did not qualify as a "site").

[73] Blaze Constr. Co., IBCA No. 2863, 91-3 B.C.A. (CCH) ¶ 24,071, at 120,506 (1991).

[74] *Id.*

[75] See §§ **8.9** and **8.10**.

[76] Simpson Constr. Co., VABCA No. 3176, 91-3 B.C.A. (CCH) ¶ 23,630 (1990).

[77] *Id.* at 118,391.

When the contract does not indicate an exact measurement or description of a condition, the analysis of materiality is more difficult. In one such case, core borings taken from an approved quarry and set forth on the contract drawings were found to "indicate" that the quarry was of reasonably good commercial quality.[78] This was found despite warnings in the contract that excessive fines, or waste rock, would be encountered.[79] In finding the amount of waste rock materially different and granting the Type I claim, the court stated that the warning rationally meant that instead of the normal amount of quarry waste, approximately 10 percent, a contractor might expect 15 or 20 percent or perhaps even somewhat more. The court went on to say that the contract could hardly mean to cover the unusual amount of over 60 percent that was actually encountered.[80] With contracts that only impliedly state a condition or estimate quality or quantity, it can be expected that a larger variance from what is actually encountered will be needed to support a finding of materiality.

§ 8.7 —Foreseeability

To be compensable, asserted subsurface or latent physical conditions encountered must not only differ materially from the contract indications, but also must be reasonably unforeseeable on the basis of all information available to the contractor at the time of bidding.[81] The condition must not have been anticipated by the contractor from a study of the contract, inspection of the site, and general experience.[82]

§ 8.8 Type II Claims

Even in the absence of specific contract representations or indications regarding any subsurface or latent physical conditions, a contractor may still be entitled to an equitable adjustment for encountering a differing site condition. Under a Type II claim, the contractor is entitled to an adjustment if it encounters an

[78] Kaiser Indus. Corp. v. United States, 340 F.2d 322, 329 (Ct. Cl. 1965) (based on specific facts of this case, the court rejected defendant's argument that "approval" of the quarry meant nothing more than permission to use the quarry, without obligation on the defendant).

[79] *Id.*

[80] *Id.* at 327. The court also held that conditions at the quarry were of an unknown and unusual nature differing materially from those ordinarily encountered. Therefore, the contractor's equitable adjustment was based on both a Type I and a Type II claim. *Id.* at 336.

[81] Mojave v. United States, 3 Cl. Ct. 353 (1983).

[82] Hannan, *What's Different about Differing Site Condition Claims?* 57 Def. Couns. J. 306 (1990).

unknown and unusual condition that is materially different from that ordinarily encountered.[83]

A Type II analysis is not concerned with representations or conditions indicated in the contract. The contract is silent as to the conditions to be encountered. The silence need not be as to all conditions at the site, but must exist at least in regard to the condition forming the basis of the Type II claim. The focus is on what was actually encountered compared to what is normally or usually encountered at a like jobsite. The contractor must prove that what it encountered was not normal but, in fact, unknown and unusual.[84] In contrast, a Type I claim focuses on a comparison between what was actually encountered and what the contract documents said would be encountered. The lack of contractual representations regarding the jobsite in a Type II case makes the contractor's burden of proof more difficult to meet.

> A Government construction contractor seeking to establish a 'category two' changed condition is confronted with a relatively heavy burden of proof. In the case of a 'category one' changed condition, the Government has, with relative precision, represented the subsurface or latent physical conditions to be encountered, and if it turns out that they have been materially misrepresented, a claim has been established. Under 'category two,' in contrast, the Government has elected not to presurvey and represent the subsurface conditions with the result that a claimant must demonstrate that he has encountered something materially different from the 'known' and the 'usual.'[85]

This is necessarily a stiffer test because of the wide variety of materials ordinarily encountered when excavating.[86] This wide variety makes it difficult for a contractor to prove what it encountered was truly unusual.

In addition to showing the encountered condition was unusual, the contractor has a "greater duty to conduct prebid inquiries or reasonable site inspections because recovery on a Type II claim is available only if a condition is unknown. A condition is not unknown if it would have been revealed upon inquiry or a reasonable site investigation."[87]

Although establishing that the differing site condition was unknown and unusual is the key to a successful Type II claim, a four-element test is employed

---

[83] Imbus Roofing Co., GSBCA No. 10430, 91-2 B.C.A. (CCH) ¶ 23,820, at 119,348 (1991).

[84] *Id.* at 119,348.

[85] Charles T. Parker Constr. Co. v. United States, 433 F.2d 771, 778 (Ct. Cl. 1970); Hallmark Elec. Contractors, Inc., ASBCA No. 32595, 87-2 B.C.A. (CCH) ¶ 19,870, at 100,493 (1987); Imbus Roofing Co., GSBCA No. 10430, 91-2 B.C.A. (CCH) ¶ 23,820, at 119,348 (1991).

[86] Charles T. Parker Constr. Co. v. United States, 433 F.2d 771, 778 (Ct. Cl. 1970).

[87] Shumate Constructors, Inc., VABCA No. 2772, 90-3 B.C.A. (CCH) ¶ 22,946, at 115,190 (1990).

when analyzing such claims. To prevail in a Type II case, a contractor must prove:

1. What were the recognized and usual conditions at the site?
2. What physical conditions were actually encountered?
3. Did the physical conditions actually encountered differ materially from the known and the usual?
4. If so, did they cause an increase in the cost of performance?[88]

These additional elements are essentially the logical extensions of proving a condition was unknown and unusual.

Several of the requirements needed for a Type II claim are identical to those required for a Type I claim. (An in-depth analysis of the factors that are unique to a Type II claim follows in §§ **8.9** and **8.10**.)

A Type II claim, just as a Type I claim, must involve (1) a physical condition (2) at the site. The analysis used to determine if these two criteria have been satisfied is the same for both a Type I and a Type II claim.[89]

Additionally, a determination of materiality, although more difficult in a Type II claim for lack of an "indication" in the contract documents, is substantially the same as the Type I analysis. The determination of whether the differing site condition is materially different from the known and usual is intricately related to a determination of whether the alleged condition is unusual in the first place. Therefore, materiality in a Type II claim, although previously addressed in § **8.6** under a Type I claim, is discussed in more depth in § **8.11**.

Although reasonable reliance is technically needed to successfully assert a Type II claim, it is rarely an issue. In a Type I claim, the contractor must prove it reasonably relied on encountering "indicated" conditions, while in a Type II claim it must show it relied on encountering "usual" conditions.[90] A contractor alleging a Type II condition generally assumes the presence of the usual condition unless it knew or should have known of the unusual condition.[91] Because of the court's recognition of this assumption, reliance is rarely an issue in a Type II claim. However, in a case based on nondisclosure, the contractor must have been misled by the nondisclosure and show it did not anticipate the existence of the differing site condition.[92]

---

[88] Charles T. Parker Constr. Co. v. United States, 433 F.2d 771, 778 (Ct. Cl. 1970); Sandwich Islands Constr., Inc., ASBCA No. 35244, 88-3 B.C.A. (CCH) ¶ 21,143, at 106,739 (1988).

[89] See § **8.3**.

[90] Illinois Constructors Corp., ENG BCA No. 5678, 91-2 B.C.A. (CCH) ¶ 23,728, at 118,846 (1991).

[91] *Id.*

[92] Leal v. United States, 149 Ct. Cl. 451 (1960); Joseph A. Cairone, Inc., ASBCA No. 20504, 81-2 B.C.A. (CCH) ¶ 15,220, at 75,367 (1981).

Thus, the primary elements of a Type II claim are whether the condition was unknown and whether it differs materially from the usual.

§ 8.9 —Unknown Condition

The language of the differing site conditions clause states the existence of a Type II condition must have been unknown to the contractor.[93] It must be a condition that reasonably could not have been anticipated by the contractor from a study of the contract documents,[94] by site investigations,[95] or by the general experience of the contractor.[96] This test is both subjective and objective. The claim will be denied if the contractor knew or should have known of the existence of the condition.[97]

Note that the unknown condition providing the basis of the claim may relate to the specific properties of a known condition. For example, the unexpectedly sound condition of 50-year-old concrete, to be removed under a dam rehabilitation contract, constituted a Type II differing site condition. The soundness of the concrete and the concomitant unsuitability of the contractually listed standard methods of concrete removal were unusual and unexpected.[98] In effect, the board was saying that although the existence of the concrete was known, its unusually sound condition was not.

Although a literal reading of site conditions clauses does not restrict recovery to conditions that existed at the time the contract was formed, the Armed Services Board of Contract Appeals (ASBCA) has interpreted such a restriction.[99] However, this does not appear to be an absolute rule. In another case, a board of

---

[93] F.A.R. 52.236-2(a)(2) (1984).

[94] Medsger, *Category II Differing Site Conditions in Construction Contracts,* Army Law., June 1988, at 12 (citing B&M Roofing & Painting Co., ASBCA No. 26998, 86-2 B.C.A. (CCH) ¶ 18,833 (1986); C&L Constr. Co., ASBCA Nos. 22993, 23040, 81-1 B.C.A. (CCH) ¶ 14,943, *aff'd on reconsideration,* 81-2 B.C.A. (CCH) ¶ 15,373 (1981); Callaway Landscape, Inc., ASBCA No. 22546, 79-2 B.C.A. (CCH) ¶ 13,971 (1979)).

[95] *Id.* (citing White Cap Painters, ASBCA No. 25364, 81-2 B.C.A. (CCH) ¶ 15,195 (1981) (site investigation revealed some paint was peeling; therefore, contractor should have anticipated same problem); Titan Midwest Constr. Corp., ASBCA No. 23594, 81-1 B.C.A. (CCH) ¶ 15,067 (1981) (contractor should have known of possible subsurface water because site was near a pond, creek, and drainage ditch); Lunseth Plumbing & Heating Co., ASBCA No. 25332, 81-1 B.C.A. (CCH) ¶ 15,063 (1981) (unusual height of tub faucets could have been determined by visual inspection)).

[96] *Id.* (citing Potomac Co., ASBCA No. 25371, 81-1 B.C.A (CCH) ¶ 14,950 (1981); Hoyer Constr. Co., ASBCA No. 21616, 84-2 B.C.A. (CCH) ¶ 17,249 (1984)).

[97] *Id.* (citing Ellis Constr. Co., ASBCA Nos. 19541, 19714, 75-1 B.C.A. (CCH) ¶ 11,238 (1975)).

[98] Illinois Constructors Corp., ENG BCA No. 5678, 91-2 B.C.A. (CCH) ¶ 23,728 (1991).

[99] Medsger, *Category II Differing Site Conditions in Construction Contracts,* Army Law., June 1988, at 15 (citing Randall H. Sharpe, ASBCA No. 22800, 79-1 B.C.A. (CCH) ¶ 13,869 (1979); Acme Missiles & Constr. Co., ASBCA No. 10784, 66-1 B.C.A. (CCH) ¶ 5418 (1966)).

contract appeals has held that sand that had been deposited uphill from the site by another contractor and washed down during heavy rain was a differing site condition.[100] However, at least one commentator has noted that this case should not have been based on the differing site conditions clause, but rather on a breach of the government's implied duty to cooperate and not hinder contract performance.[101]

In determining if the condition was unknown, one of the first questions is whether the contractor conducted a reasonable site investigation. As stated previously, a claimant asserting a Type II claim has a greater duty to conduct pre-bid inquiries or reasonable site inspections.[102]

Typically, the contract that contains a differing site conditions clause will likewise contain a site investigation clause. This is almost certain to occur in a public works contract.[103] Under such a clause, a contractor acknowledges that it has conducted a site investigation and is satisfied with the conditions. Although not explicit in the language, the site investigation clause has been interpreted by the board to constitute a waiver only for conditions that could have been discovered by a reasonable site investigation.[104]

The sufficiency of the site investigation is measured by what a reasonably experienced, intelligent, contractor-layperson should have anticipated or discovered.[105] In *Northwest Painting Service,* the contractor was unable to prove entitlement to an adjustment under a Type II claim because, as the board of contract appeals stated, the investigation that was conducted was cursory and inadequate.[106] Additionally, the contractor's failure to become aware of the soft condition of the concrete to be painted, which was substantially damaged by sandblasting, was determined to be due to the contractor's inexperience and lack of knowledge when it conducted the site inspection.[107]

As shown in *Northwest Painting Service,* an inspection by a contractor may be inadequate and therefore insufficient to justify the contractor's assertion that the condition was unknown. Likewise, a contractor will not be successful in

---

[100] Medsger, *Category II Differing Site Conditions in Construction Contracts,* Army Law., June 1988, at 10, 15 (citing Frank W. Miller Constr. Co., ASBCA No. 22347, 78-1 B.C.A. (CCH) ¶ 13,039 (1978)).

[101] *Id.* at 15.

[102] Shumate Constructors, Inc., VABCA No. 2772, 90-3 B.C.A. (CCH) ¶ 22,946, at 115,190 (1990).

[103] F.A.R. 52.236-3 (1984) (requires insertion of site investigation clause in all contracts using differing site conditions clause as expressed in F.A.R. 36.503).

[104] C&L Constr. Co., ASBCA Nos. 22993, 23040, 81-1 B.C.A. (CCH) ¶ 14,943, *aff'd on reconsideration,* 81-2 B.C.A. (CCH) ¶ 15,373 (1981).

[105] Northwest Painting Serv., Inc., ASBCA No. 27854, 84-2 B.C.A. (CCH) ¶ 17,474, at 87,076 (1984).

[106] *Id.*

[107] *Id.*

asserting a Type II claim if the condition complained of was not discovered because of a complete lack of an inspection. The clearest example would be if rock outcroppings were visible at a site. In these cases, a board of contract appeals has determined that these outcroppings gave the contractor notice that subsurface rock might be encountered once excavation began.[108] In line with the requirement that the inspection only need be reasonable, a board has held that if such outcroppings were hidden by heavy vegetation or foliage, the contractor's failure to discover them during its site investigation would be excused.[109]

A contractor need not make its own subsurface exploration, but must carefully examine any the government provides.[110]

Failing to investigate or failing to conduct a reasonable investigation will not automatically preclude recovery. The contractor only assumes the risk of differing site conditions that would have been apparent from a reasonable inspection.[111] If no inspection was conducted, the contractor has the burden of proving a reasonable inspection would not have revealed the differing site condition.[112]

§ 8.10 —Unusual Condition

Determining if a condition is "unusual" is the most troublesome element of a Type II claim. As indicated earlier, a wide variety of conditions may be encountered in any given construction project. It is not at all uncommon for the contractor to encounter a condition it did not anticipate. Because of this phenomenon, the majority of litigation regarding Type II claims centers around the proof that the condition encountered was indeed unusual.

An *unusual condition* within the meaning of a differing site conditions clause has been defined as one that might not reasonably be anticipated based on the contract work and the location at which it was to be performed.[113] A contractor is not entitled to expect the most favorable conditions, but it need not anticipate the worst.[114] One need not encounter a geological anomaly, such as permafrost in the tropics, to recover.[115]

---

[108] COVCO Haw. Corp., ASBCA No. 26901, 83-2 B.C.A. (CCH) ¶ 16,554 (1983); Overhead Elec. Co., ASBCA No. 22210, 78-2 B.C.A. (CCH) ¶ 13,440 (1978).

[109] Robert D. Carpenter, Inc., ASBCA No. 22297, 79-1 B.C.A. (CCH) ¶ 13,675 (1979).

[110] Hurlen Constr. Co., ASBCA No. 31069, 86-1 B.C.A. (CCH) ¶ 18,690 (1985).

[111] Leiden Corp., ASBCA No. 26136, 83-2 B.C.A. ¶ 16,612 (1983); Praxis-Assurance Venture, ASBCA No. 24748, 81-1 B.C.A. ¶ 15,028 (1981); AAAA Enters., Inc., ASBCA No. 28172, 86-1 B.C.A. (CCH) ¶ 18,628 (1985).

[112] Tutor-Saliba, ASBCA No. 23766, 79-2 B.C.A. (CCH) ¶ 14,137 (1979).

[113] Kinetic Builders, Inc., ASBCA No. 32627, 88-2 B.C.A. (CCH) ¶ 20,657, at 104,400 (1988).

[114] *Id.*

[115] *Id.*

The Type II claims may involve conditions occurring in nature, such as a certain rock composition,[116] or they may involve problems created by man-made conditions, such as metal debris uncovered during excavation.[117]

As previously stated, the first criteria of the four elements needed to prove a Type II claim is a showing of what the recognized and usual conditions at the worksite were.[118] Although proof of this criteria is typically the flip side of proving a condition was unusual, it should not be taken lightly. As a board of contract appeals has expressed, "absent *specific* evidence of what is typical or usual, we cannot conclude that the [claimed condition] was unusual."[119] Consequently, no matter how strange the condition the contractor encountered happens to be, the contractor will not succeed in asserting a Type II claim unless it shows what is usually encountered in like jobsites. Proof of a Type II claim requires a comparison, which necessitates the establishment of what was usual so the reviewing body has something to compare with what was actually encountered.

Contractors may use a variety of methods to show what conditions are recognized or usual and why what they encountered was not. One of the most common and successful methods involves a consideration of what is usual in that geographic area. This is typically accomplished through expert testimony.[120]

As stated earlier, the unusual condition in a Type II claim may be one occurring naturally. For example, in an area commonly made up of glacial till, expert testimony showed that encountering a pocket of cobbles in that geographic area was an anomaly.[121] Therefore, the contractor was entitled to recover for the additional cost resulting from drilling piles into the loose material.[122] Additional natural phenomena that have been found unusual in a Type II claim because of either their nature or their location include substantial quantities of muck on a jobsite at a high elevation in an area where nearby projects had not encountered muck,[123] volcanic eruption,[124] and rock with crevices or seams atypical for a given geographical area.[125]

The determination of whether an encountered condition is usual or unusual has typically been anything but scientific. Because of the variety of conditions

---

[116] See § **8.9**.

[117] See § **8.9**.

[118] See § **8.8**.

[119] Imbus Roofing Co., GSBCA No. 10430, 91-2 B.C.A. (CCH) ¶ 23,820, at 119,349 (1991); Instalaciones Aire Frio S.A., ENG BCA No. PCC-44, 85-2 B.C.A. (CCH) ¶ 18,015, at 90,352 (1985).

[120] Leiden Corp., ASBCA No. 26136, 83-2 B.C.A. (CCH) ¶ 16,612 (1983) (both the contractor and the government provided expert testimony as to the amount and type of crevices normally found in rock in the area of the jobsite).

[121] Hurlen Constr. Co., ASBCA No. 31069, 86-1 B.C.A. (CCH) ¶ 18,690, at 93,959 (1985).

[122] *Id.*

[123] Kinetic Builders, Inc., ASBCA No. 32627, 88-2 B.C.A. (CCH) ¶ 20,657 (1988).

[124] Highland Reforestation, Inc. v. United States, 230 Ct. Cl. 1020 (1982).

[125] Leiden Corp., ASBCA No. 26136, 83-2 B.C.A. (CCH) ¶ 16,612 (1983).

encountered, as well as the possible degrees of variations among conditions, the analysis of Type II claims has been very fact specific.

If the contractor is unable to show that the condition it encountered was unusual for the area, its Type II claim will be disallowed. This is precisely what happened when a contractor attempted to show the lava rock ledge it encountered when excavating was unusual for Hawaiian soil at that depth.[126]

The condition that is allegedly unusual for a given geographical area may also be a man-made condition or structure. In fact, because of the wide variety of natural conditions normally encountered in construction, Type II claims are usually more successful when dealing with a man-made condition than with one occurring in nature.

The considerations when dealing with a man-made condition are quite similar to those addressed when dealing with a natural condition. The geographical location of the structure or condition will be considered, and in the case involving a structure, the time of construction is important. In one case, a construction contractor was entitled to an equitable adjustment for removing a brick wall encountered in the interior of a hospital.[127] In determining that this was an unusual condition, the board based its decision on the contractor's unchallenged affidavits establishing that in a hospital constructed during the 1940s it would be extremely rare and unusual to find in an interior building a solid masonry wall of four layers of brick with a plaster finish.[128] The brick wall was encountered because it had originally been part of the exterior wall of the hospital before an addition had been constructed.[129]

Other man-made conditions that have been found adequate to sustain a Type II claim include the presence of jet fuel and steam in a subterranean culvert under repair[130] and a sacred Indian religious site.[131]

Another category of Type II claims that occurs quite frequently involves the presence of debris at the construction site. In one such case, a contractor was to clean out heating ducts in barracks at Fort Bragg.[132] When cleaning out the ducts, the contractor encountered substantial debris in the duct work that constantly entangled and damaged the contractor's vacuuming machines. The debris included beer cans, jars of jam, gun powder, live ammunition, ladies' underwear, and a whiskey still in one of the furnace rooms. The government

---

126 COVCO Haw. Corp., ASBCA No. 26901, 83-2 B.C.A. (CCH) ¶ 16,554 (1983).

127 Hercules Constr. Co., VABCA No. 2508, 88-2 B.C.A. (CCH) ¶ 20,527 (1988).

128 *Id.* at 103,772.

129 *Id.*

130 W.S. Meadows Eng'g, Inc., ASBCA Nos. 32536, 33271–75, 88-2 B.C.A. (CCH) ¶ 20,616 (1988).

131 Blaze Constr. Co., IBCA No. 2863, 91-3 B.C.A. (CCH) ¶ 24,071 (1991) (claim denied because location of Indian religious site was determined not to be part of construction site).

132 Community Power Suction Furnace Cleaning Co., ASBCA No. 13803, 69-2 B.C.A. (CCH) ¶ 7963 (1969).

asserted that the condition of the heating ducts was not at all unusual for military barracks. A board of contract appeals, however, found these conditions to be most unusual.

Other debris cases typically involved items uncovered during excavation. Subsurface metal debris such as snowmobiles, cars, and aircraft landing mats have been determined to not constitute a Type II differing site condition with respect to certain sites.[133] Although such debris might be unusual in other areas, the lack of neatness and the sense of clutter at the performance site placed the contractor on notice of the condition actually encountered.[134] On the other hand, a contractor was entitled to compensation for additional cost incurred in removing debris such as old railroad timbers, logs, and other foreign matter encountered when excavating a channel.[135]

The contractor's experience is also an important factor when considering a Type II claim. Even if an encountered condition does appear to be unusual, the claim will be denied when the contractor had knowledge that similar conditions had been previously encountered on similar work.[136] Additionally, a contractor's lack of experience is often used to defeat a Type II claim. This is precisely what happened when a contractor with over 100 years of experience engaged in a maintenance dredging contract in a geographical area in which it had no previous experience.[137] During the performance of the contract, unusually hard material was encountered. The contractor stated that in its vast experience, it had never encountered such material while conducting maintenance dredging. The board of contract appeals recognized that this perhaps was true. However, because the contractor's vast experience did not deal with high tides, strong currents, and glacial till that characterize the Anchorage Harbor, the fact that the contractor had never experienced such a hard substance was not conclusive.[138] In effect, the board was stating that, although the condition indeed may have been unusual for that specific contractor, it was not at all unusual for an Anchorage, Alaska, harbor.

A Type II claim was also denied in another Alaskan case in which a contractor encountered substantial subsurface man-made debris upon the commencement of excavation.[139] The board of contract appeals stated that the likelihood of encountering subsurface debris in tundra in the area of a settlement should have

---

[133] Arctic Slope, Alaska Gen./SKW Eskimos, Inc., ENG BCA No. 5023, 90-2 B.C.A. (CCH) ¶ 22,850 (1990).

[134] *Id.* at 114,787.

[135] Josun, Inc., ASBCA Nos. 88-113-4, 82-249-4, 88-2 B.C.A. (CCH) ¶ 20,590 (1988).

[136] Dawco Constr., Inc., ASBCA No. 31990, 88-2 B.C.A. (CCH) ¶ 20,606, at 104,138 (1988).

[137] Great Lakes Dredge & Dock Co., ENG BCA No. 5606, 91-1 B.C.A. (CCH) ¶ 23,613 (1990).

[138] *Id.* at 118,338.

[139] Arctic Slope, Alaska Gen./SKW Eskimos, Inc., ENG BCA No. 5023, 90-2 B.C.A. (CCH) ¶ 22,850 (1990).

been a matter of general knowledge to any contractor experienced in arctic work.[140]

§ 8.11 —Materiality

The next step for the contractor is to then show that the two items being compared are "materially" different.[141] Materiality is a fact-intensive issue highly dependent upon all facts and circumstances of the project contract documents and the project at issue.

In *Flores Drilling & Pump Co.,* the analysis demonstrated this interrelationship.[142] The contractor's expert stated that there was a 5- to 20-percent chance of hitting the type of material the contractor did when drilling a well deeper than 100 feet in an area such as the jobsite.[143] After quoting the *Webster's Dictionary* definition of *usual,* the board of contract appeals stated that, "A five to 20 percent chance [of the existence of the condition encountered was] not usual."[144] This is the closest a board of contract appeals has come to establishing any type of a bench mark to aid in the determination of materiality when comparing the usual with the unusual condition encountered.

§ 8.12 Proving the Claim

In general, to determine for purposes of a Type I claim if a difference exists between what the contract indicated and what was encountered, postclaim samples and drawings are made and compared with the pre-bid information given to the contractor. If there is no difference, there is no Type I claim.[145]

For example, in a contract for the maintenance dredging of a harbor, the board of contract appeals agreed that the discovery of "virgin" material would be a differing site condition. However, the contractor was unable to prove virgin material was encountered because the samples were not taken from the material in place, but from scows containing previously dredged material, and this rendered the samples questionable.[146]

A case in which the contractor did not meet the burden in a Type II claim dealt with a lack of experts.[147] Under the contract, the contractor was to remove

---

[140] *Id.* at 114,787.

[141] See § **8.8**.

[142] Flores Drilling & Pump Co., ASBCA No. 82-204-3, 83-1 B.C.A. (CCH) ¶ 16,200 (1982).

[143] *Id.* at 80,486.

[144] *Id.*

[145] Wayne Constr., ENG BCA No. 4942, 91-1 B.C.A. (CCH) ¶ 23,535 (1990).

[146] Great Lakes Dredge & Dock Co., ENG BCA No. 5606, 91-1 B.C.A. (CCH) ¶ 23,613 (1990).

[147] John C. Grimberg Co., ASBCA Nos. 40670–72, 91-1 B.C.A. (CCH) ¶ 23,454 (1990).

metal laths and plaster from concrete columns and then "to provide smooth, even surface prior to application of new finish."[148] The columns were uneven and honeycombed, and the contractor claimed a Type II differing site condition. The contractor and its three employees testified as to their experience, but the government said such combined experience is not determinative. No testing from independent experts as to whether the condition of the columns was unusual was presented. Only self-serving testimony from each side was brought into evidence. The board of contract appeals concluded that because only self-serving testimony on both sides existed and no experts were presented, the contractor failed to meet its burden. Similarly, in another case, a board of contract appeals concluded that "generalized testimony" is not enough to show if what was encountered was "materially different."[149]

---

[148] John C. Grimberg Co., ASBCA Nos. 40670–40672, 91-1 B.C.A. ¶ 23,454 AT 117,669 (1990).

[149] Imbus Roofing Co., GSBCA No. 10430, 91-2 B.C.A. (CCH) ¶ 23,820 (1991).

CHAPTER 9

COMMON REASONS FOR DENIAL OF CLAIMS

John Anthony Wolf
David L. Cole, Jr.

John Anthony Wolf is a partner in the law firm of Ober, Kaler, Grimes & Shriver in Baltimore, Maryland. Mr. Wolf specializes in construction contract negotiation, arbitration, and litigation in major commercial and industrial construction, and in public project matters. His practice also includes corporate, general business, and insurance coverage litigation. He is a member of the Litigation Sections of the Maryland State and American Bar Associations, and of the ABA Forum Committee on Construction. He serves on the Litigation Section Council of the Maryland State Bar Association and he has lectured on construction law and alternative dispute resolution procedures. Mr. Wolf is an honors graduate of the Washington and Lee University School of Law.

David L. Cole, Jr., is an associate in the law firm of Ober, Kaler, Grimes & Shriver in Baltimore, Maryland. Mr. Cole specializes in construction law and civil litigation. He is a member of the Litigation Sections of the Maryland State and American Bar Associations, and of the ABA Forum Committee on Construction. Mr. Cole received his B.A. degree from Washington and Lee University, and his J.D. degree from the University of Richmond School of Law.

§ 9.1 Failure to Appreciate Purpose of Differing Site Conditions Clauses

To understand why differing site conditions claims fail, one must first appreciate the purpose of differing site conditions clauses. In short, differing site conditions clauses alter the common law by shifting the risk for differing site conditions. Under common law, the contractor bore the risk of unforeseen physical site conditions affecting its work. Consequently, a prudent contractor first had to include in its price the cost of investigating subsurface soil conditions. The contractor also had to include a contingency to ensure it could still perform the job profitably if there were problems with the site not anticipated during the bidding stage.

Accordingly, institutional owners with large amounts of construction came to realize that they were receiving high bids to insure against the possibility of encountering unknown site conditions, even though such conditions actually arose on relatively few projects. Those owners realized that they could lower project costs in the long run by assuming the risk of unforeseen site conditions. This risk shifting allowed contractors to remove contingencies placed in their bids, which resulted in lower prices overall. This economic purpose has been recognized in language such as the following from the Court of Claims:

> The purpose of the changed conditions clause is thus to take at least some of the gamble on subsurface conditions out of bidding. Bidders need not weigh the cost and ease of making their own borings against the risk of encountering an adverse subsurface, and they need not consider how large a contingency should be added to the bid to cover the risk. They will have no windfalls and no disasters. The Government benefits from more accurate bidding, without inflation for risks which may not eventuate. It pays for difficult subsurface work only when it is encountered and was not indicated in the logs.[1]

Thus, differing site conditions clauses shift the risk to an owner for extra costs incurred. They create the right in the contractor—the express contract right—to claim for those costs. The clauses provide security against unanticipated conditions by the promise of an adjustment in the contract price and contract time if subsurface or latent physical conditions at the site differ materially from those mentioned in the contract or those ordinarily encountered.[2]

Because differing site conditions clauses alter the common law rule, however, strict compliance with their terms is required. As in other provisions

[1] Foster Constr. C.A. v. United States, 435 F.2d 873, 887 (Ct. Cl. 1970).

[2] For annotations on changed conditions, see Annotation, *Right of Public Contractor to Allowance of Extra Expense over What Would Have Been Necessary if Conditions Had Been Represented by the Plans and Specifications,* 76 A.L.R. 268 (1932), and Annotation, *Construction and Effect of a "Changed Conditions" Clause in a Public Works or Construction Contract,* 85 A.L.R. 2d 211 (1962).

altering common law relations, compliance with contract terms limiting risk shifting and/or prescribing compliance with procedural safeguards for owners is mandated. This is particularly true in differing site conditions claims in which the contractor is often viewed to occupy the better position in its relationship with an owner to identify and minimize risk. The risk-shifting security to a contractor provided by these clauses is, therefore, not to be construed as a "blank check." The clauses themselves typically also set forth precise procedures to be followed. Owners are afforded protection from avoidable, stale, and/or unjustified claims by other limiting clauses. It is noncompliance with the procedural safeguards commonly found in or accompanying differing site conditions clauses that causes otherwise valid claims to fail.

§ 9.2 Failure to Comply with Site Investigation Clauses

Differing site conditions clauses create a right to claim. Site investigation clauses limit that right. If a contractor is admonished by the contract documents to visit the site and become familiar with the conditions there, and the contractor unreasonably fails to avail itself of a site inspection, the contractor may very well find its claim for differing site conditions denied.[3]

Site investigation clauses provide that the contractor has investigated the site and has familiarized itself with site conditions. For instance, the applicable clause in the General Conditions of the American Institute of Architects (AIA) Document A201, 1976 edition, provides:

> 1.2.2. By executing this Contract, the Contractor represents that he has visited the site, familiarized himself with the local conditions under which the Work is to be performed, and correlated his observations with the requirements of the Contract documents.

A site investigation clause, however, does not in and of itself negate a differing site conditions clause in a contract. Instead, it limits the ambit of differing conditions to those that a reasonable investigation would not uncover. It is effective, however, insofar as it may prohibit a contractor from contending that actual conditions differed from what it anticipated, if the actual conditions are those that a reasonable investigation would have disclosed.

Specifically, the presence of a site investigation clause creates a duty for the contractor to investigate the site.[4] A failure reasonably to investigate the site

---

[3] Clark v. United States, 5 Cl. Ct. 447 (1984), *aff'd,* 770 F.2d 180 (Fed. Cir. 1985).

[4] *See, e.g.,* Mojave Enters. v. United States, 3 Cl. Ct. 353 (1983); Puget Sound Bridge & Dredging Co. v. United States, 131 Ct. Cl. 490, 130 F. Supp. 368 (1955).

defeats a differing site conditions claim.[5] A contractor with a contract containing a site investigation clause is required under the contract to conduct a reasonable investigation of the site. A reasonable investigation, however, does not require the contractor to be a trained geologist or other specialized expert, nor does it require that it hire such experts or conduct technical investigations.[6] However, the contractor will not be able to make a claim for that which it should have discovered as a reasonable contractor. The Court of Claims has stated the rule in the following way:

> In the cases arising under the modern changed conditions clause, caution continues to be observed that the duty to make an inspection of the site does not negate the changed conditions clause by putting the contractor at peril to discover hidden subsurface conditions or those beyond the limits of an inspection appropriate to the time available.
>
> The contractor is unable to rely on contract indications of the subsurface only where relatively simple inquiries might have revealed contrary conditions. When, for instance, logs in the contract show no water, as in the instant case, he need not seek out experts who may tell him of the possibility of subsurface water.[7]

A site investigation clause itself does not preclude reliance by the contractor on the information indicated in the contract documents.[8] Site investigation, however, requires common sense. To comply with the clause, a contractor must conduct an inspection of the site that, under the circumstances of the inspection, is reasonable and lends confirmation of or support to that indicated in the contract documents.[9] This is not to say that contractors should discount representations by the owner regarding the site conditions in the contract. The contractor is entitled to rely on that indicated in the contract. It need only make a reasonable inspection of the site given the available circumstances.[10] Moreover, it has been held that a failure of a contractor to conduct a site investigation when called to do so by the contract did not preclude a recovery on a claim when it was shown that a site investigation would not have revealed the true condition of the site in any event.[11]

---

[5] Clark v. United States, 5 Cl. Ct. 447 (1984), *aff'd,* 770 F.2d 180 (Fed. Cir. 1985); *William Maloney,* AGBCA No. 811051, 82-1 B.C.A. (CCH) ¶ 15,529 (1982).

[6] Morrison-Knudsen Co. v. United States, 170 Ct. Cl. 712, 345 F.2d 535 (1965); *see also* Structural Preservation Sys., Inc., 3 MSBCA ¶ 234 (1989).

[7] Foster Constr. C.A. v. United States, 435 F.2d 873, 888 (Ct. Cl. 1970).

[8] *See, e.g.,* Morrison-Knudsen Co. v. United States, 397 F.2d 826 (Ct. Cl. 1968).

[9] *Id., accord* J.F. Shea Co. v. United States, 4 Cl. Ct. 46 (1983), *aff'd,* 754 F.2d 338 (Fed. Cir. 1985).

[10] Morrison-Knudsen Co. v. United States, 397 F.2d 826, 842 (Ct. Cl. 1968).

[11] Mann Constr. Co., AGBCA No. 76109, 80-2 B.C.A. (CCH) ¶ 14,674 (1980).

The facts surrounding particular circumstances of a site investigation will determine whether a contractor will ultimately be successful on its claim. For instance, in *Appeal of Biggers Construction Co.,*[12] a contract was entered into for the laying of pipe in a trench and then backfilling the trench with either the excavated or borrow material. The contract contained a site investigation clause. The contractor chose not to attend a pre-bid site investigation tour and, therefore, was not aware of difficulties associated with working with the excavated soil. Instead, the contractor elected to perform its own site investigation, which consisted simply of a drive-by in the contractor's personal automobile without any apparent effort to inquire into the makeup of the soils that would be encountered at the site. Delays and extra costs were incurred, for which the contractor sought reimbursement on both Type I and Type II claims under the theories of changed conditions and duty to disclose based on superior knowledge.[13]

Not surprisingly, the contractor was denied a recovery because of its cavalier approach to site investigation. Regarding the Type I claim, it was determined that the contractor's investigation was inadequate because had it properly investigated, the contractor would have discovered the problematic nature of the soil later encountered. The court determined that something more than simply driving around the site in a private automobile was required. The Type II claim was also denied based on what was determined not to be a reasonable inquiry of the soil conditions on the part of the contractor.[14] The lesson to be gained from this contractor's failure to appreciate the site investigation requirements is instructive. Most cases are not quite so clear-cut.

A contractor must be careful in placing too great an emphasis on that indicated in the contract documents and not enough of an emphasis on its own site investigation. A contractor must reconcile that information provided by the contract with that found by its own site investigation. However, the indications provided by the owner in the contract may only be relied upon by a contractor if it is reasonable to do so in light of what is found in the contractor's site investigation. In *Fox v. United States,*[15] the low bidder was awarded a contract for roadwork. In the course of construction, it was necessary to perform certain excavation. When the contractor had completed, it filed a Type I claim due to encountering soil conditions that it had not anticipated based on the indications of the contract documents. It was established that the contractor had made two pre-bid site investigations.

The court denied the contractor's claim, determining (1) that the contractor unreasonably relied on the indications in the contract documents and (2) did not give enough weight to its own site investigations. The court determined that the site visits should have revealed to a reasonably prudent contractor the conditions

[12] EBCA No. 46479, 81-1 B.C.A. (CCH) ¶ 14,848 (1981).

[13] *Id.*

[14] *Id.*

[15] 7 Cl. Ct. 60 (1984), *aff'd,* 785 F.2d 326 (Fed. Cir. 1985).

encountered.[16] Because site investigation clauses are founded on common sense principles, a court may hold that a contractor with requisite experience knew (or should have known) that indications in a contract were so different from that which could be expected at the site to preclude its claim.[17]

§ 9.3 Failure to Understand Elements of Claim

The most common type of claim, termed a *Category* or *Type I* claim, involves a condition that differs materially from those indicated in the contract documents. Proof of this sort of claim essentially raises the question as to what the contract documents indicate about site conditions.

A Type I claim is not dependent on a finding that the contractor was misled by the owner as to the nature of the subsurface conditions, or that the owner misrepresented the nature of the subsurface conditions, or, finally, that the owner withheld or concealed relevant information within its grasp. In a Type I claim, fault on the part of the owner is not a necessary element. The elements to be established for a Type I claim provide a "road map" for the contractor to follow in establishing a claim. Without each of the following elements, however, a Type I claim will fail:

1. The contract documents must have affirmatively indicated or represented the subsurface conditions that form the basis of the claim
2. The contractor must act as a reasonably prudent contractor in interpreting the contract documents
3. The subsurface conditions actually encountered within the contract site area must have differed materially from the subsurface conditions indicated for that area in the contract documents
4. The contractor's claimed excess costs must be shown to be solely attributable to the materially different subsurface conditions at the contract site.[18]

The key to success on such a claim usually lies in the ability to demonstrate what the contract documents indicate about conditions. If the contract documents contain nothing, there is no recovery under a Type I claim. However, note that the word *indicates* has been interpreted broadly. In other words, the contract "indications" need not be explicit, but only enough to lead a reasonable bidder not to expect the conditions actually encountered.[19] Indications have been

---

[16] *Id.*

[17] *See* Wm. A. Smith Contracting Co. v. United States, 412 F.2d 1325, 1333–34 (Ct. Cl. 1969).

[18] *See* Hardaway Contractors, Inc., 3 MSBCA ¶ 227 (1989); Weeks Dredging & Contracting, Inc. v. United States, 13 Cl. Ct. 193, 218 (1987), *aff'd,* 861 F.2d 728 (Fed. Cir. 1988).

[19] See § 9.4.

found based upon specifications of design or construction procedures reasonably implying that a certain condition existed, even though nothing in the contract documents expressly stated that condition was present.[20]

A second type of claim, known as a *Category* or *Type II* claim, is not based upon indications in the contract documents, but upon encountering conditions (in the words of the standard federal clause) "differing materially from those ordinarily encountered and generally recognized as inhering in work of the character provided for in the contract."[21] Type II claims are generally more difficult to prove than Type I claims. A Type II claim does not involve a comparison between contract indications and actual conditions. Instead, it requires a comparison of actual conditions with those reasonably anticipated, taking into account those factors that a prudent, experienced bidder customarily uses in making judgments about site conditions, quantity, quality, and methods of performing the particular work.

One key to recovery under this type of claim lies in proving that a site investigation would not and did not disclose the condition. Evidence concerning industry custom, common knowledge, and traditional assumptions ordinarily involved in bidding upon a particular type of work will likely be introduced by the contractor to prove, or the owner to disprove, the reasonableness of the contractor's expectation as to what is "usual." The pitfalls of not understanding and, thus, not establishing these elements result in the failure of these claims.

§ 9.4 Failure to Use Data Properly

A contractor cannot expect to succeed on a Type I claim based on information in a contract that is patently unreliable.[22] Accordingly, if what is indicated by the contract documents is not pertinent to the project site, obviously incomplete, or in some other way so lacking that a reasonably prudent contractor could not justify reliance upon that indicated, a contractor's claim will be denied.[23] An exception to this general rule may lie when no other information purporting to conflict with the contract documents is available to the contractor.[24] In other words, to the extent that a recovery, if any, will be awarded to a contractor relying on clearly deficient data, the recovery will be limited to such loss that could be attributed to reasonable reliance.[25]

A significant amount of litigation has arisen over what the word *indicates* means, and the courts have interpreted the term broadly. For instance, in *Pacific*

---

[20] *See* Foster Constr. C.A. v. United States, 435 F.2d 873 (Ct. Cl. 1970).

[21] FAR 52.236-2 (1991).

[22] Morrison-Knudsen Co. v. United States, 170 Ct. Cl. 712, 345 F.2d 535 (1968).

[23] *See, e.g.*, Clark v. United States, 5 Cl. Ct. 447 (1984), *aff'd*, 770 F.2d 180 (Fed. Cir. 1985).

[24] *Id.*

[25] *Id.*

Alaska Contractors, Inc. v. United States,[26] the Court of Claims held that contract "indications" do not have to be explicit, but only enough so that a reasonable contractor would not expect that which it encountered at the site.[27] Obviously, if the contract documents do not indicate anything, a recovery by a contractor based on what is indicated in the contract documents can be expected to be rejected.[28]

To succeed on a differing site conditions claim, a contractor must be able to establish that its claim is based on reasonable interpretation of the indications in the contract documents.[29] In doing so, the contractor will need evidence of acting prudently in interpreting the contract documents and in making its bid for the project.[30] Failure to present such evidence will defeat a claim. Therefore, careful documentation should be kept of all bidding efforts, including, perhaps, site inspection, information requested, questions asked, responses given regarding the project, and pre-bid calculations resulting in the final bid.

A court, in determining whether a contractor was reasonable in interpreting that indicated in the contract, must "place itself into the shoes of a 'reasonable and prudent contractor' and decide how such a contractor would act in [the contractors's] situation."[31] It is a determination depending on the facts of each particular case. For example, if the information contained in the documents constituting the contract makes apparent that it is not to be relied upon, but nevertheless is relied upon by a contractor, the contractor may be determined not to have reasonably interpreted the indications in the contract.[32]

To establish that the date provided in the contract documents was used properly in preparing a bid, the contractor must first show that it reviewed the data in making its decision to bid on a project.[33] For instance, in *Highland Construction Co. v. Stevenson,*[34] a subcontractor began excavation work on a roadway project before the contract negotiations were complete and a formal contract was executed. During the interim period between when the subcontractor began work on the project and when the contract was finally executed, site conditions were encountered that were different from those indicated in the contract documents. The subcontractor sought recovery, but its claim was denied because it gained actual knowledge of the site conditions before the contract was executed. Accordingly, the court determined that the subcontractor could not establish

---

26 436 F.2d 461 (Ct. Cl. 1971).

27 *Id.* at 469.

28 *Id.*

29 *E.g.*, P.J. Maffei Bldg. Wrecking Corp. v. United States, 732 F.2d 913 (Fed. Cir. 1984).

30 Weeks Dredging & Contracting, Inc. v. United States, 13 Cl. Ct. 193 (1987), *aff'd,* 861 F.2d 728 (Fed. Cir. 1988).

31 P.J. Maffei Bldg. Wrecking Corp. v. United States, 732 F.2d 913, 917 (Fed. Cir. 1984).

32 *Id.*

33 Flippin Materials Co. v. United States, 160 Ct. Cl. 357, 312 F.2d 408 (1963).

34 636 P.2d 1034 (Utah 1981).

reasonable reliance on what were potentially defective plans for the construction of the project.

A contractor must also show that it reviewed and relied on the entirety of the contract documents, rather than only on certain information made available to it. In *Flippin Materials Co. v. United States,*[35] a contractor entered into a contract for the excavation of rock. The contractor based its bid on information supplied by the government, including test borings and field logs. After beginning work, the contractor claimed that the government's indications were not accurate. The court denied the claim, citing that the contractor had failed adequately to review all the contract documents because logs of test excavations existed that contained more information than what was contained in the contract drawings. According to the court, the logs were available, would have been presented to the contractor on request, and would have made apparent to the contractor the conditions that could have been expected at the site.

In like fashion, when reviewing all of the contract documents indicating site conditions, a contractor cannot simply disregard indications it considers to be unfavorable. In *J.A. Terteling & Sons, Inc. v. United States,*[36] drill hole information in the contract documents indicated both favorable and unfavorable conditions at the project site. After work began at the site, conditions were encountered leading to a claim. The Court of Claims denied the contractor's claim because it determined that the contractor failed to take into proper account all of the information in estimating its work and used only that information favorable to its cause in preparing its bid. Such selectivity equates to failure of a claim.

A contractor may also be determined not to have properly used indications in the contract documents when prior experience with similar conditions in the same locale should have alerted the contractor to conditions to be encountered. In such circumstances, a contractor may not simply rely on the information provided by the government owner in the face of known conditions.[37] The contractor may also be precluded from establishing that its use of owner-provided data was reasonable where the contractor conducted its own test borings. That was the holding in *Umpqua River Navigation Co. v. Crescent City Harbor District.*[38] There, a contractor entered into a contract for the dredging of a boat basin. A subcontractor experienced cost overruns due to soil conditions encountered in the dredging area that were allegedly inaccurately represented in the contract documents and relied upon by both the contractor and subcontractor. The court denied the claim, determining that the contractor's claim was deficient, largely because employees of the subcontractor had encountered similar conditions on

---

[35] 312 F.2d 408 (Ct. Cl. 1963).

[36] 390 F.2d 926 (Ct. Cl. 1968). *See also* Erickson-Shaver Contracting Corp. v. United States, 9 Cl. Ct. 302 (1985).

[37] Umpqua River Navigation Co. v. Crescent City Harbor Dist., 618 F.2d 588 (9th Cir. 1980).

[38] *Id.*

previous endeavors near the project site that should have, according to the court, alerted the contractor to the possibility of encountering the conditions.

The court also determined that the contractor's claim was deficient because the subcontractor conducted its own test borings even though the changed conditions clause in the general contract with the owner released the contractors from an obligation to do their own investigative digging at the site. The court determined that "[t]he fact that [the subcontractor] made its own borings indicates that it was not completely relying on the [owner's] boring results."[39] Accordingly, a contractor may be precluded from claiming that it reasonably relied on and properly used an owner's data pertaining to a project when it conducted its own tests.[40]

A contractor that anticipated the conditions encountered at the site before entering into the contract cannot later make a changed conditions claim based on indications in the contract documents. In *Wm. A. Smith Contracting Co. v. United States,*[41] a contractor entered into a contract for roadway construction in Alaska. The contract called for excavation work at specified locations along the route of a highway and for the placement of suitable excavated material as embankment and shoulders. The embankment was to consist of suitable material and was to contain no "muck." After the contractor began the work, frozen material or permafrost was encountered, and the contractor requested permission to use borrow material in the placement of the embankments instead of the excavated frozen material.

The contract drawings indicated that permafrost was present, but only at specified locations at the site. However, the terms of the contract stated that no representation was made that the conditions in the contract drawings would not be found in areas other than those shown in the drawings. At the site, the contractor found that the permafrost indicated in the contract drawings was predominant throughout the site and so made a claim. The Court of Claims denied the claim. The court determined that the fact that the contractor encountered the permafrost at locations not indicated in the contract drawings did not establish that conditions it encountered were materially different from those indicated in the contract documents. Much of the basis the court used in denying the contractor's claim was that the contractor admitted that it had anticipated encountering the permafrost before entering into the contract.

§ 9.5 Failure to Take Disclaimer Clauses Seriously

Contractors should be aware and wary of disclaimer clauses often included in construction contracts. Again, the purpose of the clause must be understood to

---

[39] *Id.* at 595.

[40] *See id.; see also* Fox v. United States, 7 Cl. Ct. 60 (1984).

[41] 412 F.2d 1325 (Ct. Cl. 1969).

appreciate the risk to a successful claims presentation. The purpose of a disclaimer clause is to allow an owner to disclaim altogether the responsibility for the authenticity or accuracy of all documents or representations purporting to represent site conditions. In essence, the clause attempts to take back that which an owner initially grants in the differing site conditions clause.

Consequently, disclaimer clauses have not had a favorable reception in federal and other public contract settings. For instance, the Court of Claims stated:

> Even unmistakable contract language in which the Government seeks to disclaim responsibility for drill hole data does not lessen the right of reliance. The decisions reject, as in conflict with the changed conditions clause, a "standard mandatory clause of broad application," the variety of such disclaimers of responsibility—that the logs are not guaranteed, not representations, that the bidder is urged to draw his own conclusions.[42]

Too often, however, contractors do not pay serious enough attention to disclaimer clauses. Accordingly, the prudent course is that if it is in the contract, believe it. It is a basic tenet of contract law that individuals are free to contract as they so choose and, accordingly, will be bound to the terms of their contract. This is particularly true in private sector contracting. Despite criticism of disclaimer clauses, they have been upheld to prevent recovery by contractors even in cases where the contractor is able to prove that the site condition information supplied to it was incorrect. Cases exist holding that disclaimers of the accuracy of the information in a contract preclude a contractor from arguing that reliance on the information was reasonable.

In *Zurn Engineers v. State,*[43] the contract clearly warned the bidders that all statements of "quantities are approximate only, being given as a basis for the comparison of bids, and that the [state] does not expressly or by implication, agree that the actual amount of work will correspond [with the estimates]."[44] California courts upheld the clause.[45] The case is illustrative of owners' attempts to eliminate the potential for liability for any possible discrepancy indicated in the contract documents and the fact that they very well may prevail.

In a private sector case, *Frontier Foundations, Inc. v. Layton Construction Co.,*[46] a contractor entered into a major contract for the construction of a building. In preparing its bid, the contractor used boring logs in the plans and specifications of the contract documents the owner provided. The contractor then contracted with a subcontractor to drive interlocked sheet piles into the earth.

---

[42] Foster Constr. C.A. v. United States, 435 F.2d 873, 888 (Ct. Cl. 1970).

[43] 69 Cal. App. 3d 798, 138 Cal. Rptr. 478, *cert. denied,* 434 U.S. 985 (1977).

[44] 69 Cal. App. 3d at 803.

[45] Zurn Eng'rs v. State, 69 Cal. App. 3d 798, 138 Cal. Rptr. 478, *cert. denied,* 434 U.S. 985 (1977).

[46] 818 P.2d 1040 (Utah 1991).

The subcontractor, in preparing its bid, used the same boring logs that the contractor had used in bidding the general contract.

The specifications described the boring logs as being from "a represented area near the site" (in fact, 1,400 feet from the site), but cautioned that they were "not part of the contract documents" and "were not a warrant of subsurface conditions."[47] The boring logs indicated that no gravel and/or cobble were encountered at the actual site tested. However, the contractor and subcontractor encountered both gravel and cobble at the project site after work began.

The contractor made a claim. The court denied the claim because the boring logs accurately represented that they pertained only to an area near the site and that they were clearly disclaimed as not being part of the contract. The Utah court, making note that it was considering a private sector and not a public works contract, determined that the contract, viewed as a whole, was not ambiguous. Accordingly, the court found that the contractor was precluded from being able to establish reasonable reliance.

Even in a federal works setting, and despite the fact that disclaimers are not favored by courts or administrative boards, if properly drafted with very explicit language, they may control.[48] Similarly, in a New Jersey public works case,[49] the state court denied a contractor's claim based on the fact that the contract explicitly stated that any information regarding conditions at the site was not to be relied on.[50] The contractor made a differing site conditions claim despite this clause. The court denied recovery, stating that the contractor could not "overcome the specific disclaimer in the specifications for any information concerning subsurface conditions."[51]

Disclaimer clauses have also been held effective in defeating claims when they state that the information provided is not to be considered part of the contract.[52] In addition, disclaimer clauses have been held to be effective when the disclaimer in the contract suffices as a release by the contractor of any right to rely on the information.[53]

Specifically, in *Joseph F. Trionfo & Sons v. Board of Education,*[54] a contractor entered into a contract for excavation work. The test boring data was not

---

[47] *Id.* at 1041.

[48] P.J. Maffei Bldg. Wrecking Corp. v. United States, 3 Cl. Ct. 482, 732 F.2d 913 (Fed. Cir. 1984).

[49] Sasso Contracting Co. v. State, 173 N.J. Super. 486, 414 A.2d 603 (App. Div. 1980).

[50] *Id.* at 491, 414 A.2d at 605.

[51] *Id.* (quoting Golomore Assocs. v. New Jersey State Highway Auth., 173 N.J. Super. 55, 413 A.2d 361 (App. Div. 1980)).

[52] North Harris County Junior College Dist. v. Fleetwood Constr. Co., 604 S.W.2d 247 (Tex. Civ. App. 1980); *accord* Metropolitan Sewerage Comm'n v. R.W. Constr., Inc., 72 Wis. 2d 365, 241 N.W.2d 371 (1976).

[53] Joseph F. Trionfo & Sons v. Board of Educ., 41 Md. App. 103, 395 A.2d 1207 (1979).

[54] *Id.*

provided with the contract specifications but was available upon request. The contractor was required to make the request for the data in the following form:

> Please forward copies of test boring data sheets for the subject project. *The contracting firm herein named releases the Owner and Architect from any responsibility* or obligation as to its accuracy or completeness or for any additional compensation for work performed under the contract due to assumptions based on use of such furnished information.[55]

The Maryland appellate court denied the contractor's subsequent claim, determining that its request for data constituted an executed release of any right to rely on the data provided.

Disclaimers may also be given effect when they recite that the indications in the contract documents do not make positive representations, but only make suggestions from which no conclusions are to be drawn.[56] Similarly, in a federal setting in *P.J. Maffei Building Wrecking Corp. v. United States,*[57] disclaimer language was enforced when the contract provided:

> Some drawings of some of the existing conditions are available for examination at [another location]. These drawings are for information only and will not be part of the contract documents. The quantity, quality, completeness, accuracy and availability of these drawings are not guaranteed. Prospective bidders shall telephone . . . for an appointment to examine drawings of the existing conditions.[58]

The Claims Court held that because of the disclaimer, the contract documents did not indicate anything. In validating the disclaimer, the court stated that any indications were "provided for information only . . . not to be part of the contract and that their quantity, quality, completeness, accuracy, and availability" were not guaranteed.[59]

Disclaimer provisions have also been given effect when they specifically limit the indications in the contract documents to specific areas.[60] For example, again in a federal setting, if that indicated in the contract pertains only to specific areas at the site, a contractor will be unsuccessful on a claim based on that which was indicated in the contract encountered at an altogether different location at the site.[61]

If the contract contains an express disclaimer clause, a contractor's ability to prove one of the few common law exceptions may also be limited. For example,

---

[55] *Id.* at 105, 395 A.2d at 1209 (emphasis added).

[56] Sasso Contracting Co. v. State, 173 N.J. Super. 486, 414 A.2d 603 (App. Div. 1980).

[57] 732 F.2d 913 (Fed. Cir. 1984).

[58] *Id.* at 914.

[59] *Id.* at 917.

[60] Wm. A. Smith Contracting Co. v. United States, 412 F.2d 1325 (Ct. Cl. 1969).

[61] *Id.*

reasonable reliance is typically an essential element to proof of a claim of fraud or negligent misrepresentation. A disclaimer clause expressly providing that the contractor should not rely on any site information, therefore, may very well present a significant obstacle to claims of fraud or misrepresentation.[62]

§ 9.6 Failure to Give Timely Notice

A contractor must establish compliance with the notice provisions in the contract to succeed on a differing site conditions claim. A failure to do so may defeat an otherwise proper claim.[63] To understand why such a draconian result may arise, the rationale for notice must also be understood. Simply stated, notice is required to allow an owner the opportunity to inquire into the accuracy of the claim and to investigate to determine the least expensive solution. It allows an owner to protect itself. The owner has accepted the risk shifting in agreeing to a differing site conditions clause, and prompt notice affords an owner immediate response and a prompt investigation when evidence is still available and fresh in the minds of participants (and potential future witnesses). Without proper notice, an owner has no way of fully defending a claim or limiting its liability.[64]

A contractor must establish that written notice of the differing conditions was provided to the contracting officer in either a Type I or Type II claim. The notice must be given before the differing conditions are disturbed. If the contractor fails to give written notice, but the government owner had actual notice of the conditions encountered, the contractor may be permitted to pursue the claim.[65] Contractors should be aware, however, that courts give heavy deference to a contract's notice provisions.

Typical adherence to the importance of timeliness is found in the Connecticut case of *Cicio Brothers v. Town of Greenwich.*[66] There, a contract was entered into for the construction of a school. A subcontractor bid on a part of the job requiring borrow fill based on the contract indications. The subcontractor's bid substantially underestimated the amount of fill required, and it claimed against the contractor and the owner. The contract required that all claims were to be submitted not more than two days after first sustaining a loss or performing

---

[62] *See, e.g.,* Eastern Tunneling Corp. v. Southgate Sanitation Dist., 487 F. Supp. 109 (D. Colo. 1979).

[63] Schnip Bldg. Co. v. United States, 645 F.2d 450 (Ct. Cl. 1981).

[64] *E.g.,* Blankenship Constr. Co. v. North Carolina State Highway Comm'n, 28 N.C. App. 550, 222 S.E.2d 452 (1976).

[65] Brinderson Corp. v. Hampton Rds. Sanitation Dist., 825 F.2d 41 (4th Cir. 1987); Roger J. Au & Son, Inc., IBCA No. 1303-9-79, 84-1 B.C.A. (CCH) ¶ 17,094 (1984).

[66] 156 Conn. 561, 244 A.2d 404 (1968).

additional work. The state court, relying on that notice requirement in denying the claim, provided the rationale underpinning notice requirements:

> Had the plaintiff given timely notice in accordance with the terms of its subcontract, the contractor and the town would have had an opportunity not only then and there to have inquired into the accuracy of the plaintiff's claims as to the fill requirements but, more importantly, the town would have had an opportunity to save the added expense and keep within the limits of the public funds appropriated for the project by changing site plans or reducing costs in other portions of the project.[67]

The admonition, "when all else fails, read the contract," perhaps is no more applicable than in notice issues, a lesson restated by the United States District Court for the Eastern District of Virginia in *McDevitt & Street Co. v. Marriott Corp.*[68] Applying Virginia law to that private sector construction dispute, and noting that the construction contract "constitutes the law that governs the parties' relationship," the court rigidly construed failure to give contractually mandated timely written notice as a complete bar to claims for additional compensation and time.[69]

In *McDevitt & Street Co.,* soil conditions caused delays in the construction of a hotel, and the project was not completed on time. The general contractor made many claims for additional costs, some related to the soil conditions, others not. Some of the claims pertaining to extra work were specifically denied because the contractor did not submit proper notice to the owner.

The contract provided that seven days' written notice was required from the contractor before proceeding on work for which the contractor would claim additional money. The contract provided that failure to give the requisite notice "deprived" the contractor of its right to claim such extra costs. Because the contract contained the written notice provision, the court held that the contractor's claim was "expressly" barred. The court noted that failure to give timely notice barred the owner from reexamining its drawings and selecting less costly alternatives.[70]

Notice requirements, however, may also be subject to an argument of waiver. When the owner knows that the work was being performed and knows that the contractor considered that the work would result in additional cost, waiver of notice may be established. The owner must have had actual knowledge of the matters that would have otherwise been in the written notice and received it

---

[67] *Id.* at 568, 244 A.2d at 407.

[68] 713 F. Supp. 906, 919 (E.D. Va. 1989), *aff'd in part, rev'd in part,* 911 F.2d 723 (4th Cir. 1990).

[69] *Id.*

[70] *Id.*

sufficiently in time to respond to a claim adequately.[71] However, a contractor should not rely too heavily on such a waiver theory. Even when the owner knows of changes, waiver may not be established. In *McDevitt & Street Co.,* for instance, the federal court rejected the contractor's argument of waiver even though it was based on the architect's being the first on the project to bring the changed condition to the contractor's attention. The court found an absence of evidence that the architect or owner was aware of the attendant costs or of any "clear and convincing" evidence of implicit waiver, and thus the contract notice requirements were not altered.[72]

Notice is more than a procedural nicety. Indeed, notice goes to the very substance of a claim. That was the reasoning of *Stauffer Construction Co. v. Board of Education.*[73] There, the Maryland appellate court was faced with a question of arbitrability of claims for compensation and time extensions for extra work, in part due to changed conditions on a rehab project. In what is clearly dicta in addressing the trial court's function in determining waiver of arbitration, the appellate court nonetheless used strong language in interpreting the 1976 AIA Document 201 General Conditions notice requirements. The court stated that noncompliance with the contract's written notice provisions for time extensions or extra compensation "operates to invalidate or annul the claim itself."[74]

§ 9.7 Failure to Consider Effects of Weather

Conditions arising from adverse weather alone and its effects, without something more, are typically not the type of changed conditions justifying a claim. The general rule is that weather conditions do not constitute a basis for a claim under a changed conditions clause. Changed conditions claims, including those based on excessive rainfall, excessive snowfall, severe wind, and sand, ice, and frozen ground, have been found not to constitute a compensable changed condition.[75]

Time extensions, independent of a changed conditions clause, however, may be granted if a delay in work is due to extremely harsh and unforeseeable

---

[71] *See, e.g.,* Singleton Contracting Corp., IBCA No. 14131280, 81-2 B.C.A. (CCH) ¶ 15,269 (1981).

[72] 713 F. Supp. 906, 919 (E.D. Va. 1990).

[73] 54 Md. App. 658, 460 A.2d 609 (1983).

[74] *Id.* at 671, 460 A.2d at 616.

[75] *E.g.,* Hardeman-Monier-Hutcherson v. United States, 458 F.2d 1364 (Ct. Cl. 1972); *see also* McCutcheon-Peterson, IBCA No. 1392980, 81-1 B.C.A. (CCH) ¶ 14,497 (1981); Praxis-Assurance Venture, ASBCA No. 24748, 81-1 B.C.A. (CCH) ¶ 15,028 (1981); Evergreen Eng'g, Inc., IBCA No. 994573, 78-2 B.C.A. (CCH) ¶ 13,226 (1978); Warren Painting Co., ASBCA No. 18456, 74-2 B.C.A. (CCH) ¶ 10,834 (1974); S.S. Mullen, Inc., ASBCA No. 8583, 65-1 B.C.A. (CCH) ¶ 4644 (1965); Overland Elec. Co., ASBCA No. 454, B.C.A. (CCH) ¶ 4359 (1964).

weather.[76] What constitutes adverse weather conditions justifying a time extension is a matter of proof that the claiming contractor must establish. Unusual weather conditions are typically proven by looking to climatological data for an extended period (such as 10 years) before the period in question and examining the frequency with which the weather conditions at issue occurred.[77]

In what is often considered the leading case on the matter, *Arundel Corp. v. United States,*[78] a contract was entered into for dredging work in Cape Cod Canal, Massachusetts. Before the work began, a severe hurricane hit, the result of which was an increase in work needed to meet the contract requirements. The contractor submitted a claim under the changed conditions clause in the contract.

The Court of Claims denied the claim, explaining that unusually bad weather is not covered by a changed conditions clause because it is a basic tenet of contract law that neither party to a contract is responsible to the other for damages brought about by such adverse weather conditions or acts of God unless the obligation has been expressly assumed in the contract.[79] In the absence of any such provision according relief, a claimant will not be entitled to recover under a changed conditions clause.

Implicit in the court's determination was that indications in a contract's documents pertain only to conditions present at the site at the time of entering into the contract. Accordingly, adverse weather conditions or acts of God that may occur in the future cannot be part of a contract's indications. The risks of such events, then, are to be assumed by the parties entering into a contract, and each party is to bear the consequences, depending upon how they fall.[80]

Similarly, in *Norfolk Dredging Co. v. United States,*[81] a contract was entered into for the construction of a levee requiring extensive fill. Progress on the job was slowed and made more difficult due to a hurricane that delayed and damaged work by increasing the contractor's fill requirements under the contract. The contractor made a changed conditions claim under the contract based on the hurricane. The court, following the general rule, stated "[a]ssuming the correctness of the finding that the storms and hurricanes were chiefly responsible for the excessive [fill requirements] which increased the plaintiff's costs, indisputably such acts of God are not considered to constitute changed conditions."[82]

---

[76] Herman H. Neumann, IBCA No. 6821167, 70-1 B.C.A. (CCH) ¶ 8216 (1970); *see also* Unitec, Inc., DOT CAB No. 7729, 78-1 B.C.A. (CCH) ¶ 12,958, *aff'd,* 78-2 B.C.A. (CCH) ¶ 13,222 (1978).

[77] *See* Construction Pan-Caribe, S.A., ENG BCA No. PCC 18, 73-2 B.C.A. (CCH) ¶ 10,238 (1973); Ray W. Lynch, IBCA No. 764269, 69-2 B.C.A. (CCH) ¶ 8026 (1969).

[78] 103 Ct. Cl. 688, *cert. denied,* 326 U.S. 752 (1945).

[79] 103 Ct. Cl. 688.

[80] *See id.*

[81] 360 F.2d 619 (Ct. Cl.), *cert. denied,* 385 U.S. 919 (1966). *See also* Arundel Corp. v. United States, 103 Ct. Cl. 688, *cert. denied,* 326 U.S. 752 (1945).

[82] Norfolk Dredging Co. v. United States, 360 F.2d 619, 634 (Ct. Cl.), *cert. denied,* 385 U.S. 919 (1966) (citing Warren Bros. Rds. Co. v. United States, 123 Ct. Cl. 48, 105 F. Supp. 826

The court also determined that no exception to the general rule could apply because the contract provided that the contractor assumed the cost of excessive fill requirements due to any causes.

The general rule has been applied in a number of unique circumstances. In *Turnkey Enterprises v. United States,*[83] a changed conditions claim based on drought was denied. In the case, a contract was entered into for certain reconstruction of a road in a national forest. To do the work, the use of water was required, and it was anticipated to come from a nearby river. An unusually dry summer caused the river to dry up, causing the contractor to stop work. The Court of Claims held that the drought constituted an act of God and, accordingly, determined that the events did not constitute a compensable claim.

A similar determination was made in *Appeal of Roen Salvage Co.,*[84] in which an extremely high water level in Lake Michigan caused the contractor to have to work at the site in considerably deeper water than anticipated. The Corps of Engineers Board of Appeals held that the unusually high water level amounted to an adverse weather condition and denied recovery.

An exception to the general rule regarding changed conditions claims due to adverse weather claims may be available if the unforeseen severe weather causes a loss when combined with some other condition that was not indicated in the contract documents.[85] For instance, in *John A. Johnson Contracting Corp. v. United States,*[86] the contractor was allowed to recover on a changed conditions claim when a severe freeze and subsequent thawing of ground conditions combined with the deficient design and construction of roads at the project site resulted in a complete disintegration of the roads, causing delays in the work.

In the case, an early spring and heavy rain caused roads leading to and from the site over which heavy hauling was being conducted to thaw. Because of the thaw, trucks sank in the mire and caused delay in the work. A factor contributing greatly to the delay was that the roads were not properly constructed in that the earth bases upon which the gravel for the road was placed were frozen when the roads were built. Drainage ditches on the sides of the roads were also inadequate. The combined effect of the thawing, early spring rains, and deficient construction of the roads contributed to poor drainage, together delaying the work.

(1952)); Breymann v. United States, 106 Ct. Cl. 367, 65 F. Supp. 398 (1946); Arundel Corp. v. United States, 103 Ct. Cl. 688, *cert. denied,* 326 U.S. 752 (1945); Arundel Corp. v. United States, 96 Ct. Cl. 77 (1942); *but see* John A. Johnson Contracting Corp. v. United States, 132 F. Supp. 698 (Ct. Cl. 1955) (contractor allowed to recover due to unforeseen severe freezing and subsequent thawing that disintegrated roads to and from site).

[83] 597 F.2d 750 (Ct. Cl. 1979).

[84] ENG BCA No. 36701, 79-2 B.C.A. (CCH) ¶ 13,882 (1979).

[85] *See, e.g.,* John A. Johnson Contracting Corp. v. United States, 132 Ct. Cl. 645, 132 F. Supp. 698 (1955); *accord* Southern Paving Corp., AGBCA No. 741031, 77-2 B.C.A. (CCH) ¶ 12,813 (1977).

[86] 132 Ct. Cl. 645, 132 F. Supp. 698 (1955).

Adverse weather conditions may also be compensable if superior knowledge regarding weather conditions vital to performing the contract is possessed by the owner but not shared with the contractor before entering into the contract. An affirmative duty to disclose such knowledge exists.[87] A failure to do so constitutes a breach of contract.[88] As noted, absent such circumstances, however, the effects of weather will generally not support a differing site conditions claim.

§ 9.8 Failure Due to Assumption of Risk

In *Cal-Pacific Foresters,*[89] a snowstorm prevented a contractor from conducting a site inspection. The contractor, nevertheless, bid on the project and was awarded the contract. The Department of Agriculture Board of Contract Appeals determined that the contractor was precluded from making a differing site conditions claim because it assumed the risk of whatever conditions it might encounter at the site.

The case involved a contract with the United States Forest Service for the planting of trees. A tour of the site was to be conducted with the forest service, but a snowstorm caused the site inspection to be canceled. The contractor, nevertheless, bid on and was awarded the contract. When it began work, the contractor sustained losses due to the slope and makeup of the soil at the site and submitted a claim based on a changed condition.

The board denied the contractor's claim. Although the evidence of what was indicated to the contractor about the conditions of the site was in conflict, the board of appeals determined that the conditions at the site were "latent or unknown to the contractor because he was unable to view the planting area before he submitted his bid . . . [and] in doing so without full knowledge of the terrain undertook a venture the risk of which rested on his shoulders rather than the Government's."[90] Without expressly stating, the board of appeals denied the contractor's claim based on the doctrine of "assumption of risk."[91]

§ 9.9 Failure to Consider Local Ordinances

Changes to local ordinances affecting the site where the work is to be performed have been found not to constitute a changed condition justifying a claim.

---

[87] Hardeman-Monier-Hutcherson v. United States, 458 F.2d 1364 (Cl. Ct. 1972); *accord* Utility Contractors, Inc. v. United States, 8 Cl. Ct. 42 (1985), *aff'd,* 790 F.2d 90 (Fed. Cir.), *cert. denied,* 479 U.S. 827 (1986).

[88] Hardeman-Monier-Hutcherson v. United States, 458 F.2d 1364, 1372 (Cl. Ct. 1972).

[89] AGBCA No. 230, 70-1 B.C.A. (CCH) ¶ 8087 (1970).

[90] *Id.*

[91] *See id.*

In *Edwards v. United States,*[92] a United States Postal Services contract was entered into for the construction of a post office building. The postal service implemented an *open advertising* method of procuring a location for the building, in which developer bidders own or control the proposed site. Such a method is different from a so-called *controlled site* method, in which the government owns and hence controls the site. When using the open advertising method, the government typically does not investigate zoning and other ordinances pertaining to a proposed site.

The contractor submitted its bid but did not inquire into present or pending zoning or local ordinances applicable to the site. Instead, the contractor relied solely on general representations made to it by a postal service real estate specialist who himself had not properly investigated the issue. After entering into the contract, zoning regulations and local ordinances, some ratified after the execution of the contract, required the contractor to perform unanticipated additional work. For example, an ordinance adopted after the contract was executed required the contractor to improve streets not originally contemplated in its bid.

The contractor submitted a changed conditions claim. The Claims Court determined that the claim was not recoverable because the contractor had a duty to investigate, which it failed to fulfill. According to the court, an investigation conducted by the contractor would not have been overly burdensome and would have revealed the present or pending zoning regulations and local ordinances. Moreover, the court determined that the contractor's reliance on the postal service's real estate specialist was not reasonable. The court also pointed out that the contract documents imposed upon the contractor the duty to investigate zoning regulations. Finally, the court noted that the contractor was responsible under the contract to comply with zoning regulations or local ordinances whether they became effective before or during the term of the contract.[93] Again, the lesson is that if it is in the contract, believe it.

§ 9.10 Failure to Document and Prove Entitlement and Damages

Although differing site conditions clauses create a contractual right to recovery, enforcing the contractual right is an altogether different matter. It is not a simplistic process, and a contractor must have evidence to support its claim. The burden of proving a site conditions claim is borne by the contractor,[94] and before

---

[92] 19 Cl. Ct. 663 (1990).

[93] *Id.*

[94] J.F. Shea Co. v. United States, 4 Cl. Ct. 46 (1983), *aff'd,* 754 F.2d 338 (Fed. Cir. 1985).

a contractor will be granted recovery, it must establish that it is in compliance with all contractual prerequisites to establish a claim.[95]

Once the contractor establishes that it is entitled to a differing site conditions recovery, it still must establish damages or equitable adjustment. As in all construction claims, adequate recordkeeping is essential. All too often, however, contractors focus almost entirely on establishing entitlement and fail to document and establish resultant costs and damage. Accordingly, information that should be meticulously documented may include, but not be limited to, descriptions of the changed conditions encountered, activities, specific areas at the site where the changed conditions were encountered, notices given, actions taken in mitigation, and a contemporaneous detailed listing of all direct costs because of or related to the changed condition.

---

[95] *E.g.,* S.J. Groves & Sons Co. v. State, 50 N.C. App. 1, 273 S.E.2d 465 (1980), *reh'g denied,* 302 N.C. 396, 279 S.E.2d 353 (1981).

CHAPTER 10

PROVING AND PRICING DAMAGES

Bruce R. Darroch
Richard J. Sieracki

Bruce R. Darroch is vice president and manager of the Phoenix, Arizona, office of Peterson Consulting L.P. He has extensive experience in all aspects of litigation support, including construction and government contracts. Mr. Darroch received his B.A. in economics from Kalamazoo College in Michigan, and his B.S. in Civil Engineering from the University of Michigan. He is a member of the American Society of Civil Engineers, American Association of Cost Engineers, and Construction Financial Management Association.

Richard J. Sieracki is vice president of Peterson Consulting L.P., specializing in management services in the construction and utility fields. He received a B.S. in civil engineering from the University of Illinois. Mr. Sieracki has been responsible for the preparation and analysis of many complex claim matters throughout his career. He is a member of the American Association of Cost Engineers and the Project Management Institute.

§ 10.1 Introduction

A frequent dispute on construction projects involves differing site conditions. A differing site condition occurs when the contract documents and a site investigation do not anticipate the site conditions the contractor actually encounters. For example, the drawings issued on a project indicate that the site is primarily composed of sand. The contractor confirms that the work will only involve excavation of sand by performing a reasonable investigation of the site. Upon starting work, the contractor discovers a strata of rock. Thus, a differing site condition has been encountered.

Most contracts anticipate the occurrence of differing site conditions and contain language intended to address these types of variations. However, despite the proactive attempts of owners and their representatives to draft contractual language that will limit differing site condition disputes, claim situations do occur.

Proving a differing site condition and computing the associated damages requires knowledge of finance, accounting, and construction claim methods. An understanding of the available project documentation is also required to successfully prove and price a differing site condition claim.

§ 10.2 Materiality

Differing site conditions are encountered on various construction projects. Most contracts anticipate this potential situation and contain a clause within their general conditions intended to address the issue.[1] Although the typical clause may define a differing site condition, it usually does not address the materiality or significance of the differing site condition. Unfortunately, there is no standard definition addressing the issue of materiality, and it will usually be determined by the facts and circumstances of each project.

Clauses in contracts may specify that materiality is determined by the architect or other parties, such as the project engineer or construction manager. Regardless of the parties involved, disputes over materiality are likely because of required judgments and because each site tends to be unique.

Materiality, in some instances, has been determined based upon the percentage of variance between the quantities of a material anticipated and the actual quantities encountered. In one case, the Federal Board of Contract Appeals held that a quantity variation is not material until it differs by 25 percent of the original quantity. The board stated that 25 percent is an appropriate and reasonable indicator of a material variation. The board also stated that this percentage is a rule of thumb that would not necessarily apply in all cases.[2]

The fact that the 25 percent does not apply in all cases is supported by a Postal Service Board of Contract Appeals decision. In this case, a contractor

---

[1] *See* AIA A201-1987, General Conditions of the Contract for Construction 12 (14th ed.), and federal contract clauses.

[2] Yukon Constr. Corp., FAACAP No. 66-4, 65-2 B.C.A. (CCH) ¶ 5005 (1965).

had encountered a differing site condition in furnishing 6,607 cubic yards of fill versus an anticipated 5,676 cubic yards of fill, a difference of 16 percent. The board judged the 16-percent difference to be material. In coming to this decision, the board relied upon testimony that a contractor should anticipate as much as a 10-percent overrun on excavation when doing quantity take-offs. The contractor's recovery for the differing site condition was adjusted accordingly.[3]

Contract clauses such as the standard American Institute of Architects (AIA) General Conditions clause leave the issue of materiality open to judgment. Thus, the facts and circumstances of each particular case may determine materiality. This standard contractual language, however, can be modified in an attempt to minimize the potential for disputes by defining materiality on a proactive basis.

§ 10.3 Quantity Variations

When preparing a differing site condition claim, it is important to understand how the condition affects the project. A common result of differing site conditions is a variation in quantities. One quantity issue that may lead to a dispute is a difference between the quantity of material anticipated and the quantity of material actually installed or excavated. Another example of a quantity-related dispute occurs when the contract documents and an associated site investigation did not anticipate any quantities of a certain material that was subsequently encountered during the actual performance.

As stated in **§ 10.2**, materiality is a key element in determining if the changed condition would lead to an adjustment in the contract price. The standard AIA General Conditions clause addresses quantity variations, but leaves to judgment the determination of whether or not the contract price should be adjusted.[4] The judgments in the standard AIA clause are twofold. First, it must be determined that the quantity change causes "substantial inequity" to one of the contracting parties. Second, the equitable adjustment must be determined.

Disputes over quantity variances may be limited by proactively refining the standard general conditions clause. For example, language may be added to a unit price contract to state that the quantities are to be considered estimates only, and that a variation in the final quantities of plus or minus 20 percent from the estimate will not be a basis for a change in the unit prices quoted.

A variation of this proactive clause can also be developed for a firm fixed-price contract. The owner or contracting agent may request firm, fixed unit prices for any changes in quantities as compared to those encompassed by the

---

[3] Yadkin, Inc., PSBCA No. 2051, 89-2 B.C.A. (CCH) ¶ 21,709 (1989).

[4] AIA A201-1987, General Conditions of the Contract for Construction ¶ 7.1.4.

bid drawings. Separate unit prices may be developed for various types of site materials (such as sand, rock, or clay). Attempting to make this list of materials inclusive may help to minimize the potential for disputes. However, it would be virtually impossible to state all possible conditions that may be encountered and expect a contractor to quote prescriptive unit prices for each condition.

Another means of minimizing quantity variation disputes is to specify in the contract how the unit prices will be adjusted if quantities change. By agreeing to this unit price adjustment at the beginning of the job, the project participants can reduce the time spent negotiating when the differing site condition is recognized. An equitable adjustment that allows the project to proceed on its intended schedule should be the common goal of all parties encountering a differing site condition. The following example demonstrates the advantage of this type of proactive contract language.

A contract was planned for the installation of piling for a coal-fired power plant. A series of test piles had been driven and, based on the results, specifications and other contract documents were prepared. A pre-bid meeting was held, and all potential bidders were afforded an opportunity to perform an investigation of the site. Subsequently, a fixed-price contract with a unit price specified for any added piling was entered into between the owner and the selected contractor.

Shortly after the piling installation began, the project team realized that the actual subsurface conditions were different than those portrayed in the bid documents and those that could have been reasonably anticipated by the contractor. A series of tests were performed to determine if a new subsurface profile could be defined. While installation was occurring, the owner, construction manager, and contractor worked together to achieve acceptable technical and pricing solutions. These pricing discussions proceeded quickly because the unit price for additional piling in the contract was used as a starting point. The discussions only needed to focus primarily on the additional quantities required, and the project's progress was not significantly impacted.

§ 10.4 Proving Differing Site Conditions

One key to successfully proving differing site conditions is to have the proper documentation available from each stage of the project. The first stage of the project is the *bidding period,* which spans the time from the identification of the project through the initial bid solicitation and concludes when the bids are submitted. The second stage, which is referred to as the *nonimpacted period,* covers the period from contract award until the differing site condition is recognized. Depending upon the nature of the differing site condition, this second stage may end at virtually any time during the project. In fact, the differing site condition may occur at the beginning of the project, resulting in the virtual lack of a nonimpacted period. The third stage is the period during which the differing site condition is encountered. The third stage is referred to as the *impacted period* of construction. Each individual stage has special requirements to consider in attempting to maintain the appropriate level of documentation.

Bidding Period

The contractor must perform a multitude of tasks during the bidding period, all under tight time constraints. Although there may be pressure to take shortcuts, the contractor should keep in mind that proper handling of the bid documents directly relates to the recovery under a Type I condition claim. Type I claims can be classified as those in which the site conditions differ from those indicated in the bid or contract documents.

A number of tasks are usually performed during the bidding period that directly relate to successfully proving a differing site condition. The first is the contractor's review of the bid package. This bid package is usually prepared by the owner or owner's representative, who can be the architect, engineer, or construction manager. The bid package will typically consist of the general conditions, special conditions, technical specifications, and contract drawings. The contractor must carefully review each of these documents to fully understand the scope of work to be performed. The contractor can then translate these contractual requirements into a bid for the work on either a firm fixed-price basis or a unit price basis, depending on the types of pricing requested. Careful documentation of the bid-estimating process and preservation of that documentation provides a solid base to assist in proving what the contractor had included in the bid. A listing of bid documents might include:

- Bid notes and information for bidders
- Bid set (drawings and specifications)
- Other pre-bid information clarifications and addenda (bid meeting minutes, schedule)
- Bid estimate and backup (quantity takeoffs)
- Bid evaluations
- Site investigation notes, drawings, pictures, and videos.

The following example demonstrates how bid documents can be used to support a claim for differing site conditions. An earth-moving contract was issued as part of an overall construction program for a new highway. The work involved cutting and filling 500,000 cubic yards of earth over a distance of three miles. The type of soil to be moved varied from sandy clay to sandy gravel. The contractor encountered a differing site condition after 100,000 cubic yards of earth had been cut and filled. The contractor discovered that the sandy clay soil was interspersed with significant quantities of rock. This rock was not suitable for use in the areas that were to be filled with excavated material.

The contractor had factored the anticipated soil composition of sandy clay and gravel into account in preparing the bid. The contractor estimated the unit prices for the work by considering the amount of earth to be moved, the distances involved, cycle times, haul routes, machinery requirements, compaction requirements, and dust control measures, as well as the allotted time to perform the contract. In addition, the estimates reflected the contractor's historical performance in similar situations.

The differing site condition required the contractor to perform additional work to screen the unsuitable rock from the suitable fill. Accordingly, the contractor incurred substantial extra costs due to both the labor inefficiencies and the addition of the screening operation. The labor inefficiencies increased the labor costs and resulted in higher overhead due to project delay, while the screening operation increased the costs of equipment and associated operators.

An important aspect in evaluating the additional costs is to make sure that the original bid documents did not anticipate encountering any rock. In this instance, the contractor was able to provide documentation showing exactly how sufficient monies for all of the costs relating to the original scope of work had been included, and that the bid matched the bid documents provided by the owner.

Another significant event that occurs during the bidding period is the site investigation. Proper site investigations will provide the contractor with firsthand knowledge of the physical condition of the job site where the work will be performed. A reasonable site investigation can assist the contractor in answering important questions that may influence the way in which the bid is prepared. A site investigation is usually required for the contractor to preserve the right to make a differing site condition claim.

Site investigation documentation may consist of notes or comments relating to the relevant areas of the jobsite on a project drawing. It may also include notes of conversations with the engineer, architect, construction manager, or owner's representative. Pictures or videotapes of the site may also be taken, and these can be important records of the condition of the area before work begins.

Nonimpacted Period

During the second stage of the project, the nonimpacted period, the contractor can proceed in a logical manner, performing the work in the method and sequence that was originally anticipated. The documentation produced by the project participants during this stage may provide information needed to assist in proving the differing site condition and associated impacts. Examples of such documents are:

Contract	Correspondence and meeting minutes
Schedules	Engineering design records

Purchasing and material control documents	Progress records
Quality control and test reports	Cost records
Labor information	Closeout documentation

Following is a more detailed breakdown of selected construction documents:

Contract Documents

Executed contract documents
Notice of award and notice to proceed
Subcontract and supplier agreements
Revisions
- Change orders
- Field changes
- Disputed change backup

Correspondence and Meeting Minutes

General correspondence
Interoffice memos
Daily reports
Daily diaries
Project meeting minutes
Delay notices
Time extension and change order requests
Telephone call records

Schedules

Pre-bid, original, or "accepted" schedules
Master project schedules
Engineering schedules
Detailed construction schedules
Schedule update reports and logic diagrams
Schedule status reports and meeting minutes
Subcontractor schedules
Vendor delivery schedules
Performance analysis ("S" curves, productivity)
"As-built" schedules

Engineering Design Records

Shop drawings and logs (submittals and approvals)
Design drawing (including revisions)
As-built drawings
Request for information
Field sketches and directives
Field inspection reports
Coordination drawings

Purchasing and Material Control Documents

Purchase orders and vendor contacts
Quotation records (written and verbal)
Expediting reports
Delivery tickets, logs, and inspection reports
Payment invoices
Inventory records
Noncompliance notices
Lease/rental agreements

Progress Records

Daily and weekly progress reports
General project status reports
Progress photographs and videotapes
Meeting minutes
Quantity installation reports
Equipment logs
Staffing curves
Weather records
Foreman, field supervision reports, and diaries
Field rework and extra work reports

Quality Control and Test Reports

Inspection reports
Laboratory test records

Field test reports
Stop work notices
Punch lists
Field survey books
Quality assurance/quality control reports
Nonconformance reports
Safety records

Cost Records

Payment requests, payments, and backup
Contractor's estimate/budget and work sheets
Completion estimates
Extra work backup
Accounting records and cost reports
Financial statements
Payroll registers and timesheets
Internal audit reports
Time and material tickets
Invoices (material, equipment, subcontractors)
Home office expense documentation
Loan applications, agreements, and draws

Labor Information

Trade union agreements
Logs of labor stoppages
Training and test certifications
Productivity analyses
Craft level estimates
Daily craft reports

Closeout Documentation

Punch lists
Operation testing reports
Inspection and certifications

Certificates of
- Need, occupancy
- Substantial completion
- Final completion

Waivers of lien
Warranties and guarantees

Miscellaneous Records

Permits and licenses
Field diaries and construction logs
Insurance certifications/bonds
Organization charts
Telephone call records
Claims documents
Special reports (regulatory agency and government)
Internal contractor status reports

Project documents from this nonimpacted period may provide the baseline information needed to analyze the contractor's performance before encountering the differing site condition. Evaluation of this nonimpacted period is important because it may provide a period of performance that can be compared to the estimated performance developed in the bid. Differences in the contractor's performance may be evaluated to determine whether or not they are indicative of possible bid omissions or errors, such as underestimates of anticipated productivity rates. Depending upon the method used in preparing the claim, adjustments to the bid may be required.

There are also instances when a differing site condition occurs at the beginning of construction. In these situations, additional analysis may be required to determine the baseline to compare to the impacted circumstances.

Impacted Period

Documentation of the third stage, or impacted period of the contract, starts with the recognition of the differing site condition and continues until its resolution. Most contracts require formal notice upon discovery of a differing site condition. However, notice in another manner (such as through other project documents to identify alternative notice or recognition of the condition) may be possible. Notice information may be found in the correspondence and meeting minutes files. Additionally, the progress records can provide substantiating information on the differing site condition. Progress photographs can identify the time that the differing site condition was first observed, as well as show the extent of the differing site condition.

The documents produced by the project participants during the impacted period are similar to those from the nonimpacted period. Nevertheless, the task of proving the differing site condition may be easier if specific project documentation is modified to focus on the differing site condition. For example, the project progress schedules may be expanded to include more detailed progress reports on the impacted activities. Similarly, cost accounts may be added to provide more itemized information on the dollars associated with impacted work areas. Project performance documentation may be tracked more closely to assist in segregation between differing site conditions and other factors affecting the project.

Documentation on the impacted cost areas will be especially important for the quantification of damages. This type of documentation will also be useful in addressing the cost impact while the work is ongoing. For instance, documentation of the change in labor productivity due to the differing site condition may be used as part of the overall pricing negotiations.

§ 10.5 Pricing Considerations

Differing site conditions can have significant adverse financial impacts on all parties involved in a project. If the differing site condition leads to an overall delay in project completion, the owner may incur additional financing costs, experience lost income from potential revenues, and suffer other delay losses. Similar financial implications exist for the contractor, who may achieve less-than-planned productivity rates, incur additional overhead costs both at the jobsite and in the home office, have extended equipment rental, and have other types of cost increases.

Accurately quantifying the damages incurred by a contractor, owner, or other participant on a project affected by differing site conditions can be complex. For example, project accounting and control records may not segregate impact type costs from other costs. Also, isolating the impact of the differing site condition from other impacts such as adverse weather may be difficult. **Sections 10.6** and **10.7** discuss several contractor cost elements that may be affected by differing site conditions and the different methods that have been used to quantify the impacts to contractors. The appropriate method depends on the facts and circumstances of the project.

§ 10.6 —Cost Elements

Regardless of the method used to quantify damages, the contractor's recorded costs should be examined. In some cases, the cost accounting system may yield data that can be directly used in the damage quantification. In other instances, the data can be used in conjunction with other construction records to perform damage quantifications.

Costs associated with differing site conditions claims are often segregated into two general categories. The first category is direct costs or direct expenses. *Direct costs* can be defined as all items of expense directly incurred by or attributable to a specific assignment or task. Some of the types of costs that may be categorized as direct costs are construction labor, supervision, materials, equipment, and subcontracts. The second category of costs is *indirect* or *overhead costs,* defined as contractor costs that cannot properly and accurately be allocated to a specific operation on a project. Overhead costs may include field or site overhead and home office overhead. Examples of overhead costs are rent, supplies, taxes, utilities, and officers' salaries. Because contractors frequently define costs differently, an understanding of the nature of the costs and how various terms are used on a particular project is important.

Construction labor costs. Construction labor productivity may be viewed as a measure of time required for installation of a given quantity of work. It is often expressed as man-hours per unit, and in excavation/fill projects as man-hours per cubic yard of material excavated or installed. Labor productivity may also be expressed as the quantity of installation per unit of time (such as cubic yards per hour). To arrive at a bid price for the direct construction labor portion of a project, the contractor should estimate both the number of units to be worked and the anticipated rate at which those units can be worked, yielding the total direct construction man-hours anticipated for that portion of the work. The contractor may then determine a wage rate and apply it to the direct construction man-hours to calculate the direct labor cost. This wage rate may be affected by factors such as the time period when the work will be performed, the anticipated labor mix (such as the ratio of journeymen to apprentices and foremen), and the anticipated amount of overtime.

Compared to the contractor's bid, increased labor costs may be incurred because the bid productivity rate was not achieved and additional hours had to be expended on the job. In addition, increased wages may have been incurred for premium, overtime, or shiftwork pay. Segregating the portion of this difference that may be attributable to the differing site condition from any portion that may be attributable to a difference in assumptions used in preparing the bid can be difficult. Thus, it is essential to understand the parameters and assumptions used in the bid preparation to accurately establish the amount of construction labor damages incurred.

Material costs. Differing site conditions claims may include material costs. Recorded material costs can differ from the bid estimate because of changes in quantities, prices, or the type of material used. Assessing material cost impacts requires identification of which changes occurred. Review of material invoices may be necessary to ensure that the recorded costs relate to the specific job in question and to segregate the material costs into the components of quantities and prices. In addition, the bid detail should be reviewed and compared to the recorded cost components to highlight any significant differences.

Once the significant differences are identified, the underlying reasons should be determined. For instance, a differing site condition may cause project delay and, as a result, a certain material may be purchased later than originally anticipated at a higher price. Researching project documentation and cost records can assist in distinguishing when the material costs were incurred.

Equipment costs. Equipment costs may include both rented and owned equipment. Individual contractors may assign these costs to jobs differently. In addition to understanding the contractor's job cost system, the use of each piece of equipment and its respective time on the job may need to be verified through equipment logs, daily reports, and other project documentation.

Rental equipment costs may be based upon invoices plus markups for maintenance; fuel, oil, and grease; administration; and profit. These markups should be reviewed for reasonableness and accuracy. The recorded rental equipment costs can then be compared to the bid to identify any impacts that may be due to a differing site condition.

Contractor-owned equipment costs may be charged to the project based upon hours of usage multiplied by a predetermined hourly rate. This rate would likely include the costs of ownership; maintenance; fuel, oil, and grease; and insurance. Before quantifying damages, this rate should be reviewed to confirm that it is a reasonable representation of actual costs. Recorded equipment usage and charges per hour should be compared to the bid to identify changes that may be due to differing site conditions.

Subcontract costs. Differing site conditions may impact the contractor through increased subcontractor costs. These costs may be the result of direct or ripple impacts on the subcontractor's performance. *Direct impacts* occur when the differing site condition has a primary effect on the subcontractor's cost of performance. Indirect effects may be classified as *ripple impacts.* These impacts may occur throughout the project but are still a result of the differing site conditions (for example, trade stacking (multiple trade groups working in a particular area at the same time), acceleration, or out-of-sequence work).

Identifying the increased costs begins with a review of the original subcontract to determine the anticipated scope of work. This identification process continues with an understanding of why the subcontractor's performance was altered. In most instances, the subcontractor will have asked the contractor for a change order to compensate for the differing site condition. This may necessitate a review of the change order's supporting documentation and other project documentation to analyze and assess the propriety of the change order. The recorded subcontract costs may also need to be reviewed to verify the accuracy of the recorded costs.

Field overhead costs. Increased field overhead costs may be incurred when a differing site condition impacts the project schedule or results in a change in the site support needed. Field overhead costs may include job trailers, security, water, temporary electricity, or any other indirect cost incurred on the jobsite. Analysis of field office overhead costs should ensure that any identified costs have not also been claimed as direct costs or home office overhead costs.

Increases in field overhead costs can be segregated into excess costs and extended costs. *Excess field overhead costs* are the additional costs incurred when the level of site support increases. For instance, the differing site condition may require acceleration to regain the lost schedule, and the contractor may bring additional supervisory labor on site to manage the increased work activities. *Extended field overhead costs* are the costs incurred to support site activities during the extended project period. For instance, with project delays the same level of support activities may be required, but for a longer duration than anticipated. Once the field overhead costs are established, a daily rate may be determined and multiplied by the days of delay the differing site condition caused to compute extended field overhead costs.

Home office overhead costs. Home office overhead costs typically arise from activities undertaken at an off-site location to support the various projects of the company. Examples of home office functions may include company management, accounting, bidding and estimating, contract administration, and purchasing. The home office overhead costs can be examined to determine the nature and reasonableness of such costs.

A differing site condition may cause an increase in certain components of home office overhead costs to support specific functions. Also, a differing site condition can impact the time-related home office overhead costs through project extensions. Different methods of calculating extended home office overhead damages have been used. One such method is the *Eichleay formula.* The Eichleay formula allocates home office overhead costs incurred during the contract period (generally based upon the proportions of specific job billings to total billings). Once the allocation is completed, a daily rate is calculated and multiplied by the differing site condition days of delay to arrive at damages. Modifications of the Eichleay formula have also been developed and used.[5]

§ 10.7 —Damage Quantification

Parties to a construction project have used many methods to quantify damages. The *total cost* and the *modified total cost* approach compare total actual costs incurred to estimated costs. The *cause-and-effect method* identifies the cost

---

[5] *See* R. Cushman, Proving and Pricing Construction Claims 124–25 (John Wiley & Sons 1990).

impact to specific causes on a detailed basis. This is sometimes called the *measured-mile approach* and compares actual experience for work affected by the differing site condition to the experience for similar work unaffected by the differing site condition. Other methods have also been used. Most methods compare actual contractor costs to the costs that would have been incurred had the differing site condition not existed.

Total cost method. One method that has been used to calculate damages in differing site conditions claims is the *total cost method,* in which the contractor computes the difference between actual job costs and the bid costs to arrive at the damage quantification. The bid costs used in the calculation may first be adjusted to reflect change orders and approved quantity variations.

On the surface, this damage calculation method appears to have several advantages. The method is sometimes used because it can be easily understood, is simple to prepare, and may result in the maximization of the claim amount. However, the total cost method can have serious flaws. One of these flaws is a failure to establish a cause-and-effect relationship between the changed site conditions and the actual costs of the project. There is generally no attempt to distinguish the actual costs expended in remedying the differing site condition from other costs incurred in performing the originally planned work and other changed work. Therefore, there is an inherent assumption that any cost overrun is attributable to the differing site condition and that the contractor's bid estimate would have been achieved for all other work items. The total cost method also assumes that the contractor's bid was appropriate (there were no bid errors or omissions) and that the cost of all unforeseen events or activities are the full responsibility of the defending party and are compensable.

The total cost method is therefore not appropriate unless the contractor bid can be shown to be appropriate, all contractor costs are reasonable, the contractor is not responsible for any of the increased costs, and there is no other method available for quantifying damages.

Modified total cost method. A second method, referred to as a *modified total cost method,* retains some of the simplicity of the total cost method while minimizing some of the shortfalls of the total cost method. The modified total cost method can be presented in different ways. It may focus solely on the affected aspects of the work or on the total project. For example, only the direct labor accounts affected by the differing site condition may be included in the direct cost aspect of such a claim.

To use the modified total cost method, the contractor still needs to demonstrate that the costs claimed are compensable. For example, the contractor's bid must be shown to be appropriate. To be entitled to compensation, the contractor may be required to show that the actual costs are reasonable and that the increases in costs are attributable to the owner and/or some other party, for which the contractor is entitled to compensation.

Measured-mile approach. Another method to compute damages for differing site conditions that may be appropriate depending upon the circumstances is sometimes referred to as the *measured-mile approach.* This approach is a more specific method of calculating damages than the total cost methods. In determining the direct cost damages, this method seeks to establish a cause-and-effect relationship between the differing site condition and the resultant impact.

The measured-mile method requires an isolation of time periods between when the project was and was not impacted by the differing site condition. For example, once the nonimpacted and impacted periods have been determined, the productivity rate for each of the time periods can be calculated. The difference in the two productivity rates is the increase in man-hours per installed unit that may be attributable to the differing site condition. This difference is then multiplied by the quantities installed to arrive at the direct man-hours associated with the claim. Appropriate overheads and other indirect costs are then added to arrive at the total damages.

The application of this measured-mile approach attempts to demonstrate what the contractor's productivity would have been had the differing site condition not occurred. This method, unlike the total cost and modified cost methods, does not rely on the contractor's bid estimate as a baseline. Rather, the analysis looks at the actual performance history on nonimpacted accounts for the particular project in question to establish a baseline.

There may still remain significant challenges in the acceptance of this method depending upon the case facts and circumstances. As discussed, the measured-mile approach requires an isolation of the time periods between when the impact of the differing site condition was being experienced as compared to when the condition was not experienced. The project facts and records may not lead to a clear isolation of these two periods. In addition, this method requires the isolation of the impact of the differing site condition from other impacts that may be occurring. If, for example, the nonimpacted productivity rate or baseline is established during a time period of good weather, and the period impacted by the differing site condition is affected by adverse weather, the resultant productivity now has two contributing events that may be difficult to separate. Therefore, as with any damage approach, the quantitative analysis must be linked with the analysis of project records, documents, and other information to identify all of the facts that may have impacted the contractor's performance.

A modification to this approach may be applied when the entire project is impacted by the differing site condition and a nonimpacted baseline period cannot be established from the existing project. Under this modified approach, the "should-have-been" or baseline productivity is determined by engineering estimates and/or by comparison to the actual productivity achieved on comparable projects. One of the challenges with this application is to demonstrate that the two projects are comparable.

Other methods. Other methods have also been used, depending on the facts of the particular matter. The *quantum meruit method* attempts to take into account the value of the benefit received by the owner in the work performed by the contractor, rather than just focusing on the costs the contractor incurred.

One of the other methods used to calculate damages is the *jury verdict method.* This method of awarding damages is used by the federal boards and courts when the amount of damages cannot be determined with certainty and the best the contractor can do is to make a reasonable approximation. The jury or board than makes a decision on damages based on this estimate. The contractor must still present proof of the nature and extent of increased costs and submit a reasonable estimate of damages based upon the information available.

Another method that a contractor may attempt to employ is one in which separate cost accounts are established to capture the man-hours incurred when working on the portion of the work related to the differing site condition. This method attempts to capture and isolate the actual costs incurred in performing the work. If the proper diligence is applied in monitoring the charging to these accounts, it may in fact serve that intended purpose. However, in practice, the ability to perform such segregation may be impractical or impossible, and also, once the actual costs have been captured, the task of determining the should-have-been cost for performing the work still exists. Thus, one may still be required to refer back to the measured mile or some other approach to arrive at a baseline for computing damages.

CHAPTER 11

HAZARDOUS WASTE DISPOSAL AND CLEANUP CONTRACTING

Michael R. Charness
Eric M. Drattell

Michael R. Charness is a partner with the law office of Howrey & Simon in Washington, D.C., where he specializes in government contract and construction claims and litigation. Mr. Charness received his undergraduate degree from the University of Pennsylvania and his law degree from Georgetown University in Washington, D.C. Mr. Charness is a member of the District of Columbia Bar and is active in a number of professional associations involving government contract and construction law. Mr. Charness also lectures and writes extensively on government contract, environmental, and construction issues.

Eric M. Drattell is a senior associate in the Washington, D.C., law firm of Howrey & Simon, where he specializes in government contract law, with a particular emphasis on construction and environmental cleanup contracting. He received his undergraduate degree from the University of Maryland and his law degree from the American University, Washington College of Law. Mr. Drattell is a member of the District of Columbia and Virginia bars, and has practiced before federal and state courts and boards of contract appeals. He also lectures and writes frequently on environmental, construction, and government contracts matters.

§ 11.1 Introduction

The scope of hazardous waste disposal and cleanup work is often ill defined at the time of contracting. Consequently, the contracting parties (and the owner's engineer) must be aware of their respective rights and obligations under the contract's differing site conditions and site inspection clauses. These clauses allocate responsibility for determining site conditions and liability if the site conditions differ from those described in the contract or ordinarily encountered and are intended to allow the contractor to bid confidently without having to include contingency factors in its bid for unknown conditions.[1]

Fortunately, the vast body of construction law has largely been interpreted to cover hazardous waste removal and cleanup contracting. Therefore, even though there are few cases specifically addressing differing site conditions in hazardous waste removal and cleanup contracting, there should be a high degree of predictability in the outcome of such cases.

This chapter addresses the hazardous waste removal and cleanup contracting environment, the differing site conditions and site inspection clauses typically found in public and private contracts, and typical differing site conditions encountered in hazardous waste removal and cleanup contracting. Finally, this chapter discusses one of the largest differing site conditions in hazardous waste removal and cleanup contracting to date—the Rocky Mountain Arsenal Basin F cleanup contract.

§ 11.2 Contracting Environment

It has often been said that the 1990s is the decade of the environment. With the Cold War now over, the federal government is reallocating a significant proportion of its defense-related resources to clean up facilities (including bases and nuclear weapons plants) contaminated by the detritus of 50 years of nuclear and chemical weapons development as well as simple benign neglect. Similarly, as a result of the enactment in 1976 of the Resource Conservation and Recovery Act (RCRA),[2] and in 1980 of the Comprehensive Environmental Response, Compensation, and Liability Act (CERCLA)[3] (which created what is commonly known as the Superfund),[4] the federal government, as well as state and local governments and private parties, are contracting to have cleanup work

---

[1] *See* Shank-Artukovich v. United States, 13 Cl. Ct. 346 (1987), *aff'd,* 848 F.2d 1245 (Fed. Cir. 1988); Alps Constr. Corp., ASBCA No. 16,966, 73-2 B.C.A. (CCH) ¶ 10,309 (1973).

[2] Pub. L. No. 94-580, 90 Stat. 2795 (1976) (codified as amended at 42 U.S.C. § 6901 (1988)).

[3] Pub. L. No. 96-510, 94 Stat. 2767 (1980) (codified as amended at 42 U.S.C. §§ 9601–9675 (1988)).

[4] *Superfund* derives its name from the fund created by a tax on the chemical and petroleum industries. 26 U.S.C. § 9507 (1988).

performed at such diverse sites as office buildings, dumps, manufacturing plants, and warehouses.

The federal government has been the primary sponsor of environmental cleanup activities. The estimates of the federal government's total investment in hazardous waste removal and cleanup work in the foreseeable future change daily. For example, the Defense Department's budget for cleanup programs has increased tenfold since 1984 and threefold since 1989. The budget, however, is probably just a starting point.[5] As illustrated by the Rocky Mountain Arsenal Basin F cleanup contract,[6] the magnitude of a hazardous waste removal and cleanup contract often increases dramatically after the work has begun when the actual site conditions are uncovered.

Under CERCLA, the Environmental Protection Agency (EPA) is the federal government agency primarily responsible for Superfund contracting.[7] The United States Army Corps of Engineers (the Corps), however, has undertaken responsibility for managing Superfund contracts with a price of more than $5 million. In addition, the Air Force (through the Center for Excellence), the Navy (through the Naval Facilities Engineering Command), and the Department of Energy (DOE) have all awarded contracts for environmental cleanup work.

Both RCRA and CERCLA provide that when the federal government undertakes to have a contaminated site cleaned, it can charge certain parties (known as potentially responsible parties or PRPs), including those that intentionally or negligently generated the hazardous waste necessitating the cleanup effort, with the costs of the cleanup.[8] Both RCRA and CERCLA, however, specifically give private parties the ability voluntarily to cleanup Superfund sites.[9] Consequently, an ever-increasing proportion of cleanup contracts is being awarded by private parties rather than the government.

Contracts awarded by federal government agencies are governed by the Federal Acquisition Regulation (FAR).[10] Each agency publishes its own FAR supplement, such as the Defense FAR Supplement (DFARS), EPA Acquisition Regulation (EPAAR), and DOE Acquisition Regulation (DEAR). The FAR and FAR supplements prescribe clauses to be inserted in contracts depending upon, among other things, the type of work to be performed and the method by which the government will pay the contractor (such as cost-plus or fixed price).[11] In

---

[5] The Congressional Office of Technology Assessment recently estimated that the federal government would spend over $500 billion on Superfund cleanup work, excluding Department of Energy sites.

[6] See **§ 11.8**.

[7] 42 U.S.C. § 9606 (1988).

[8] *Id.* §§ 6928, 9607.

[9] *Id.* §§ 6928, 9622.

[10] 48 C.F.R. § ____ (1991).

[11] *See, e.g.,* FAR pt. 52 (1992).

most cases, contracts prescribed for construction contracts are used in environmental cleanup contracts.

Contracts awarded by private parties are typically governed by either the laws of the state where the contract was made or performed. Private contracts for environmental cleanup work are often, but not always, prepared on standard forms published by the American Institute of Architects (AIA) for the use of the construction industry, including architects and engineers, construction managers, and contractors.

§ 11.3 Contract Clauses

The FAR prescribes standard clauses to be inserted in construction contracts that are also commonly used in hazardous waste removal and cleanup contracts. Of those clauses, two are of particular concern when determining the parties' rights and obligations in the event of differing site conditions.

The first clause, FAR 52.236-2, "Differing Site Conditions" (April 1984), sets forth the contractor's obligations upon discovery of a differing site condition and specifies the relief to which the contractor will be entitled if the conditions result in a change in the cost of performance:

(a) The Contractor shall promptly, and before the conditions are disturbed, give a written notice to the Contracting Officer of (1) subsurface or latent physical conditions at the site which differ materially from those indicated in this contract, or (2) unknown physical conditions at the site, of an unusual nature, which differ materially from those ordinarily encountered and generally recognized as inhering in work of the character provided for in the contract.

(b) The Contracting Officer shall investigate the site conditions promptly after receiving the notice. If the conditions do materially so differ and cause an increase or decrease in the Contractor's cost of, or time required for, performing any part of the work under this contract, whether changed or not changed as a result of the conditions, an equitable adjustment shall be made under this clause, and the contract modified in writing accordingly.

(c) No request by the Contractor for an equitable adjustment to the contract under this clause shall be allowed, unless the Contractor has given the written notice required; *provided,* that the time prescribed in (a) above for giving notice may be extended by the Contracting Officer.

(d) No request by the Contractor for an equitable adjustment in the contract for differing site conditions shall be allowed if made after final payment under this contract.

Timely notice of a differing site conditions claim, as the clause states, is crucial to recovery.[12] However, recovery has been allowed for differing site

---

[12] *See, e.g.*, J. Parr Constr. & Design, Inc. v. United States, 24 Cl. Ct. 228 (1991); Schnip Bldg. Co. v. United States, 645 F.2d 950 (Ct. Cl. 1981).

conditions when the contractor has substantially complied with the notice requirements.[13] Moreover, recovery is routinely allowed when, though the contractor did not give timely notice of the differing site condition, the government had actual knowledge and was consequently not prejudiced by the untimely notice.[14]

The second clause, FAR 52.236-3, "Site Investigation and Conditions Affecting the Work" (April 1984), shifts the burden of certain differing site conditions to the contractor by requiring the contractor to inspect the site where work will be performed and ascertain whether there are conditions that could affect the cost of performing the work:

(a) The Contractor acknowledges that it has taken steps reasonably necessary to ascertain the nature and location of the work, and that it has investigated and satisfied itself as to the general and local conditions which can affect the work or its cost, including but not limited to (1) conditions bearing upon transportation, disposal, handling, and storage of materials; (2) the availability of labor, water, electric power, and roads; (3) uncertainties of weather, river stages, tides, or similar physical conditions at the site; (4) the conformation and conditions of the ground; and (5) the character of equipment and facilities needed preliminary to and during work performance. The Contractor also acknowledges that it has satisfied itself as to the character, quality, and quantity of surface and subsurface materials or obstacles to be encountered insofar as this information is reasonably ascertainable from an inspection of the site, including all exploratory work done by the Government, as well as from drawings and specifications made a part of this contract. Any failure of the Contractor to take the actions described and acknowledged in this paragraph will not relieve the Contractor from responsibility for estimating properly the difficulty and cost of successfully performing the work, or for proceeding to successfully perform the work without additional expense to the Government.

(b) The Government assumes no responsibility for any conclusions or interpretations made by the Contractor based on the information made available by the Government. Nor does the Government assume responsibility for any understanding reached or representations made concerning conditions which can affect the work by any of its officers or agents before the execution of this contract, unless that understanding or representation is expressely stated in this contract.

Despite its broad language, the site inspection clause has been regularly interpreted to require the contractor only to bear the risk of site conditions that

[13] *See, e.g.,* General Casualty Co. of Am. v. United States, 127 F. Supp. 805 (Ct. Cl.), *cert. denied,* 349 U.S. 938 (1955); Holloway Constr. Co., ENG BCA No. 4805, 89-2 B.C.A. (CCH) ¶ 21,713 (1989).

[14] Claim allowed because no prejudice to the government: *see, e.g.,* Delphcon/J.A.M.E., J.V., PSBCA No. 1455, 88-3 B.C.A. (CCH) ¶ 21,107 (1988); Perini Corp., ENG BCA No. 3745, 78-1 B.C.A. (CCH) ¶ 13,191 (1978); Schouten Constr. Co., DOT CAB No. 77-4, 79-1 B.C.A. (CCH) ¶ 13,630 (1978); CSH Contractors, Inc., NASA BCA No. 476-2, 77-2 B.C.A. (CCH) ¶ 12,814 (1977). Claim barred because prejudice to the government: *see, e.g.,* Saturn Constr. Co., ASBCA No. 15,704 (1982); Joseph Penner, GSBCA No. 4647, 80-2 B.C.A. (CCH) ¶ 14,604 (1980).

would have been revealed by a reasonable or simple inspection.[15] Therefore, if the contractor fails to conduct an adequate site inspection, that is, one that a reasonable, intelligent contractor possessing experience in the field of work involved would be expected to discover, the contractor will be precluded from recovering for a differing site condition.[16]

Recently, the Court of Appeals for the Federal Circuit addressed the contractor's right to recover for differing site conditions under a hazardous waste removal contract in *Petrochem Services, Inc. v. United States.*[17] There, the court held that the government was obligated to pay the contractor for removing oil in excess of the contractor's estimate of the quantity when the government failed to disclose its superior knowledge about the quantity of oil to be removed, and the contractor conducted a reasonable inspection of the work site:

> The record reveals that the government knew that over 21,000 gallons spilled into the containment area and required cleanup. The government possessed a vital fact that affected the performance costs of the contract and appellant undertook to fulfill the contract absent these specific and essential elements. These costs would be reflected in any contractor's bids.[18]

When private contracts employ the same or virtually the same language as found in the site investigation and differing site conditions clauses, the parties may be able to rely upon the decisions by the government contracts boards and courts interpreting the clauses' requirements. For example, in *Brinderson Corp. v. Hampton Roads Sanitation District,*[19] the court held that the boards' and courts' interpretations of the FAR differing site conditions clause controlled when that same clause was incorporated in the parties' private contract:

> When a standardized provision widely used in construction contracts receives consistent judicial and administrative interpretation, it acquires a gloss that lends color to the words. When the same words are incorporated in later contracts, it may be presumed that the intention of the parties conforms to the earlier, consistent judicial and administrative interpretation.[20]

---

[15] *See, e.g.,* Foster Constr. C.A. v. United States, 435 F.2d 873 (Ct. Cl. 1970); J. Lawson Jones Constr. Co., ENG BCA No. 4363, 86-1 B.C.A. (CCH) ¶ 18,719 (1986); Tri-Messine Constr. Co., PSBCA No. 312, 78-1 B.C.A. (CCH) ¶ 13,023 (1978); Warren Painting Co., ASBCA No. 18456, 74-2 B.C.A. (CCH) ¶ 10,834 (1974).

[16] *See* Stock & Grove, Inc. v. United States, 493 F.2d 629 (Ct. Cl. 1974); Commercial Mechanical Contractors, Inc., ASBCA No. 25695, 83-2 B.C.A. (CCH) ¶ 16,768 (1983); Granite Constr. Co., IBCA No. 1500-8-81, 82-1 B.C.A. (CCH) ¶ 15,631 (1982).

[17] 837 F.2d 1076 (Fed. Cir. 1988).

[18] *Id.* at 1079. Note that though the contractor brought a claim under the contract's changes clause, the contractor could just as well have sought relief under the differing site conditions clause.

[19] 825 F.2d 41 (4th Cir. 1987).

[20] *Id.* at 44.

The AIA standard construction contract, AIA Document A201 (1987), contains differing site conditions and site inspection clauses similar to the FAR clauses cited.[21]

Clause 4.3.6, "Claims for Concealed or Unknown Conditions," contains notice and relief provisions similar to those found in its FAR cousin. The most significant difference between the AIA and FAR differing site conditions clauses is the time within which notice is required. Under the FAR provision, the contractor must give prompt notice, though the contracting officer has the discretion to extend that time. Under the AIA contract, prompt notice is also required, but in no event later than 21 days after the contractor discovers the differing site condition.

AIA Document A201 also contains a site inspection clause, Clause 1.2.2, "Site Investigation," which, like the FAR site inspection clause, limits the owner's liability for differing site conditions that the contractor could have discovered with a reasonable inspection.[22]

Contractors have used this clause to shift liability for differing site conditions to the owner when the contractor has made a reasonable pre-bid inspection or when such conditions would not have been discoverable upon a reasonable inspection.[23]

Frequently, an owner (including the federal government) will attempt to shift more of the risk for differing site conditions to the contractor by including in the contract a clause relieving it of all liability for subsurface conditions. In federal government contracts, the boards and courts often hold that the disclaimer is ineffective if the agency has not obtained a waiver of the required differing site conditions and site inspection clauses.[24] In private contracts, a disclaimer of liability for differing site conditions is likelier to be upheld, unless the owner materially misrepresented the site conditions.[25]

Virtually all federal government hazardous waste removal and cleanup contracts[26] include the required FAR differing site conditions and site inspection clauses. As is evidenced by the handful of cases reported to date,[27] the

---

[21] A similar clause is found in the standard construction contract prepared by the Engineer's Joint Contract Documents Committee (EJCDC).

[22] *See also* AIA A201-1987, Clause 3.2, "Review of Contract Documents and Field Conditions by Contractors," which obligates the contractor to compare field measurements and conditions to those described in the contract documents and known to the contractor.

[23] *See, e.g.,* Dravo Corp. v. Seattle, 484 P.2d 399 (Wash. 1971).

[24] *See, e.g.,* United Contractors v. United States, 368 F.2d 585 (Ct. Cl. 1966). FAR 36.502 and .503 require the contracting officer to insert the differing site conditions and site inspection clauses in all construction and demolition contracts.

[25] *See* P.T.& L. Constr. Co. v. State, 531 A.2d 1330 (N.J. 1987).

[26] Including federal grant contracts.

[27] *See, e.g.,* Ogden Allied Servs. Corp. v. General Servs. Admin. GSBCA Nos. 9916, 10,572, 1992 GSBCA LEXIS 200 (June 4, 1992); R.T. Crowley, Inc., GSBCA No. 11080 (9521)-REIN, 92-1 B.C.A. (CCH) ¶ 24,499 (1991).

hazardous waste removal and cleanup contractor will, no doubt, be held to the same standards as the construction contractor with respect to differing site conditions. Likewise, unless public policy intervenes, the private hazardous waste removal and cleanup contractor whose contract includes standard construction contract language will also likely be held to the same standards as the private construction contractor.

§ 11.4 What Is a Differing Site Condition?

As both the FAR and AIA clauses suggest, a differing site condition is a condition, including latent surface or subsurface physical conditions, that could affect the cost of and time for performing the work.[28] As discussed in §§ **11.5** and **11.6**, there are essentially two categories of differing site conditions: Type I, conditions differing from contract indications, and Type II, conditions differing from those that are ordinarily encountered and typically found in work of the character provided for in the contract. The type of differing site condition encountered may affect the contractor's recovery.

§ 11.5 Type I Conditions

A contractor may recover for a Type I differing site condition when the actual conditions differ from express or implied site conditions in the contract documents.

Clearly, if the contract makes an express representation that differs materially from actual conditions (assuming the contractor has fulfilled its site inspection obligations), the contractor will be entitled to an equitable adjustment under the contract's differing site conditions clause provided it relied on the contract's representation.[29] However, even in the absence of an express representation, the contractor may be entitled to recover for a Type I differing site condition.

For example, the contractor may be entitled to recover for a differing site condition when site conditions differ from inferences reasonably drawn from

---

[28] Notably, when the contractor has incurred increased labor costs because the government awarded another large contract in the vicinity, the contractor may obtain relief under the contract's changes clause and not the differing site conditions clause. *See* Bateson-Stolte, Inc. v. United States, 172 F. Supp. 454 (Ct. Cl. 1959).

[29] *See, e.g.*, Sanders Constr. Co., 220 Ct. Cl. 639 (1979), *rev'g* ENG BCA No. 3482, 75-2 B.C.A. (CCH) ¶ 11,481 (1975); City of Columbia v. Paul N. Howard Co., 707 F.2d 338 (8th Cir.), *cert. denied,* 464 U.S. 893 (1983); Husman Bros., DOT CAB No. 71-15, 73-1 B.C.A. (CCH) ¶ 9889 (1973); Norair Eng'g Corp., GSBCA Nos. 2394, 2630, 72-1 B.C.A. (CCH) ¶ 9305 (1972).

the contract documents,[30] or the owner has superior knowledge of site conditions that would affect the cost of or time for performance of the work.[31]

As discussed in further detail in **§ 11.8**, possibly one of the largest differing site conditions in hazardous waste removal and cleanup contracting was encountered by EBASCO at the Rocky Mountain Arsenal. There, the contract documents specified an estimated quantity of hazardous waste EBASCO was to remove and treat. During performance, however, EBASCO discovered a second, unknown layer of the waste nearly equal in volume to the first layer. Under the contract's differing site conditions clause, EBASCO and the Corps successfully negotiated an equitable adjustment to the contract price and time for performance.

§ 11.6 Type II Conditions

A contractor may recover for a Type II differing site condition when the actual conditions differ from the contractor's reasonable expectations of the site conditions, including natural and man-made conditions.

To recover for a Type II differing site condition, the contractor generally must demonstrate what it expected the physical conditions at the site to be, what it actually encountered, and that the actual conditions were unknown and unforeseeable notwithstanding a reasonable site inspection and local custom.[32] Therefore, if the contractor should have reasonably expected to encounter the site condition about which it complains, the contractor will not be entitled to relief under the differing site conditions clause.[33]

---

[30] *See* Foster Constr. C.A. v. United States, 435 F.2d 873, 878–94 (Ct. Cl. 1970) (Type I condition found when contract described dry excavation technique and contractor encountered subsurface water rendering dry excavation impossible); Western Contracting Corp., ENG BCA Nos. 3835, 3920, 4016, 4018, 82-1 B.C.A. (CCH) ¶ 15,486, at 76,805–6 (1981) (contractor reasonably inferred from suitable equipment list that it would encounter certain conditions).

[31] *See* Petrochem Servs., Inc. v. United States, 837 F.2d 1076, 1078–79 (Fed. Cir. 1988) (government failed to disclose its estimates of quantity of hazardous waste to be removed); Hardeman-Monier-Hutcherson v. United States, 458 F.2d 1364, 1371–72 (Ct. Cl. 1972) (government possessed critical facts unknown or unvailable to contractor); City of Indianapolis v. Twin Lakes Enter., Inc., 568 N.E.2d 1073, 1076–77 (Ind. Ct. App. 1991) (owner failed to disclose obstructions).

[32] *See, e.g.,* Dawco Constr., Inc., ASBCA No. 31990, 88-2 B.C.A. (CCH) ¶ 20,606 (1988), *aff'd,* 867 F.2d 615 (Fed. Cir. 1989); Chance Constr. Co., AGBCA No. 81-226-1, 84-3 B.C.A. (CCH) ¶ 17,603 (1984); A.S. Horner Constr. Co., ASBCA No. 5334, 59-2 B.C.A. (CCH) ¶ 2321 (1959).

[33] *See* Dawco Constr., Inc., ASBCA No. 31990, 88-2 B.C.A. (CCH) ¶ 20,606 (1988), *aff'd,* 867 F.2d 615 (Fed. Cir. 1989); Diamond Pac., NASA BCA No. 45-0391, 92-1 B.C.A. (CCH) ¶ 24,615 (1991).

It should be apparent that although the Type II differing site condition is rarer and more difficult to prove than a Type I differing site condition,[34] the Type II condition is likely to be a common occurrence in hazardous waste removal and cleanup contracting. Contractors frequently encounter Type II differing site conditions in, among other things, asbestos abatement, inactive waste site cleanup, and weapons complex cleanup contracts, in which the magnitude of the cleanup effort is often not fully known before contracting.

§ 11.7 Relief Available

Both the FAR and AIA differing site conditions clauses provide for the contract price or time for performance, or both, to be "equitably adjusted"[35] if the contractor encounters a compensable differing site conditions.

In federal government contracts, a contractor who encounters a compensable differing site condition is entitled to an equitable adjustment in both time and cost of performance. Therefore, the government is obligated to compensate the contractor for additional costs incurred as a result of the differing site condition, such as labor,[36] materials,[37] equipment,[38] sampling,[39] extended overhead and other delay-related costs, in addition to profit.[40] In addition, the contractor is entitled to recover the increased costs of performing other work under the contract not directly affected by the differing site condition.[41] Likewise, if the differing site condition results in a lower cost of performance, the government will be entitled to a reduction in contract price.[42]

Virtually identical remedies are available under private and nonfederal government contracts employing differing site conditions clauses similar to the FAR and AIA clauses. In *Kenny Construction Co. v. Metropolitan Sanitary District,*[43] because it had to employ a different construction method than planned,

---

[34] *See* R. Bednar, Construction Contracting 592 (1991); G. Keating & W. Gammon, Changes and Claims in Government Construction 136 (1987).

[35] FAR pt. 52, ¶ 31,346.06 (1992); AIA A201-1987.

[36] *See, e.g.,* State Mechanical Corp., VABCA No. 2797, 91-2 B.C.A. (CCH) ¶ 23,830 (1991); Darwin Constr. Co., ASBCA No. 27596, 86-1 B.C.A. (CCH) ¶ 18,645 (1985).

[37] *See, e.g.,* S. Kane & Sons, Inc., VACAB No. 1254, 78-1 B.C.A. (CCH) ¶ 13,100 (1978).

[38] *See, e.g.,* Utley-James, Inc. v. United States, 14 Cl. Ct. 804 (1988), *aff'g* GSBCA No. 5370, 85-1 B.C.A. (CCH) ¶ 17,816 (1989).

[39] *See, e.g.,* Frank Lill & Son, Inc., ASBCA No. 35774, 88-3 B.C.A. (CCH) ¶ 20,880 (1988).

[40] *See* Bruce Constr. Corp. v. United States, 324 F.2d 516 (Ct. Cl. 1963).

[41] *See* R. Bednar, Construction Contracting 613 (1991).

[42] *See, e.g.,* AFGO Eng'g Corp. v. United States, 227 Cl. Ct. 730 (1981), *aff'g* VACAB No. 1236, 79-2 B.C.A. (CCH) ¶ 13,900 (1979); Perini Corp. v. United States, 381 F.2d 403 (Ct. Cl. 1967).

[43] 309 N.E.2d 221 (Ill. 1974).

the contractor recovered costs for "removing trash, grouting, repaving the surface, trucking, trailer setup, and clearing and cleaning the site."[44]

§ 11.8 Rocky Mountain Arsenal Basin F Cleanup

In 1987, the United States Army Corps of Engineers awarded contracts for the cleanup of one of the worst hazardous waste sites in the United States, the Rocky Mountain Arsenal outside of Denver, Colorado. As reported in the February 11, 1988, edition of Engineering News Record:

> Toxic materials were disposed of at the base—often in open pits and lagoons—by the Army during the 40 years it produced munitions and chemical weapons there. Shell [Oil Co.] had leased portions of the sprawling arsenal site for the production of agricultural pesticides. Toxic compounds from both operations leached into the groundwater and migrated off-site into some private drinking water wells and public drinking water systems.

The cleanup contract called for the contractor, EBASCO Corporation, to remove approximately six million gallons of toxic liquid and contaminated soil from the 93-acre Basin F, solidify it, and deposit the concoction in a pyramid-shaped pile lined and capped with synthetic membranes.

During performance, the contractor determined that the Corps grossly underestimated the quantity of toxic liquid to remove. Because the contract's specifications had made affirmative representations about the quantity of toxic material to be removed and treated, the contractor was entitled to relief under the contract's differing site conditions clause for a Type I differing site condition.

The scope of the differing site conditions experienced in Basin F, although not common in hazardous waste removal and cleanup contracts, is sure to be found on a lesser scale in other projects. Often, as in the case of Basin F, there are few records documenting the precise type and quantity of hazardous materials disposed of at the site. Consequently, this creates a tension between the owner's desire to have the work performed within its budget or funding limitations and the contractor's desire to be paid for the work performed. Owners must be aware that if they employ risk-shifting clauses designed to insulate them from liability for differing site conditions, the contractor will likely include contingencies in its bid or may attempt to proceed against the owner for the differing site condition under a misrepresentation theory. Contractors must be aware of their obligation to conduct site inspections and to bid the work based upon the contract documents and the results of its investigation. In any event, although there may not be many differing site conditions on the scale of Basin F, there will surely be many hazardous waste removal and cleanup contracts with extensive differing site conditions.

---

[44] *Id.* at 223.

CHAPTER 12

ASSESSING POTENTIAL DAMAGES FROM ENVIRONMENTAL EXPOSURE

Michael P. Emmert
Martha Wilke Murray

Michael P. Emmert is a vice president in the Chicago office of Peterson Consulting Limited Partnership. He has over 15 years experience in the construction and environmental industries including construction claims, commercial litigation, project management and hazardous waste remediation action. Mike received his Bachelor of Science degree with a major in accounting from Illinois State University and is a Certified Public Accountant in the State of Illinois. He has authored and spoken on contract management and economic issues.

Martha Wilke Murray is an executive vice president of Peterson Consulting. She consults clients on a variety of complex business and litigation related matters in the areas of environmental, insurance, toxic tort, information management, and damages analyses. She received her Bachelor of Business Administration degree in Accounting from the University of Wisconsin, Madison. She is a Certified Public Accountant in Illinois and is a member of the American Institute of Certified Public Accountants, the Illinois CPA Society and the New York Institute of CPA's.

§ 12.1 Introduction

The discovery of hazardous waste at the site of a construction project has the potential for substantially disrupting the construction process and dramatically increasing the cost of any such project. Likewise, the discovery of different toxic wastes and greater volumes of contaminants can substantially increase the cost of a hazardous waste site remediation or cleanup project. As such situations develop, the members of a project team begin the process of identifying the mechanisms available to them to recover from the responsible organization the increased costs that have been or may be incurred. For most members of a project team, the contract for the work they are to perform may be the primary vehicle available to them for recovery. In addition, the owner, developer, and possibly others may look to other potentially responsible parties (under the federal environmental regulations) or their insurance carriers for reimbursement of these types of costs.

The claims made by parties in these situations need to be well constructed, adequately detailed, and thoroughly prepared. Each project team member should have an understanding of the approaches for preparing such claims as well as assessing those claims when presented.

Discoveries of unknown environmental hazards could likely be considered by the members of the project team as differing site conditions. The existence of such conditions at a construction project typically result in one or more of the members of the project team making claims for increased costs—costs that were incurred as a result of continuing with the project in light of the "new understanding" of the site's character. Of course, in these situations, the site or the conditions of the site have not really changed; merely the parties' knowledge about the site has been refined or expanded. Whether that knowledge was or should have been known by the parties is relevant in assessing potential differing site condition claims. Also, whether the parties considered such risks in their bidding and contracting phase of the project is also quite relevant in assessing and measuring claims of potential damages from changed site conditions.

Assessing the potential economic harm that a party may attempt to recover from others on the project team is a multistep process that begins with the determination of whether the potential claimant has standing to bring such claims. Decisions regarding claim methodology, scope of the economic harm to be measured, and the adequacy of the documentation to support a claim must be made. The ultimate decision that must be made is whether the identification and measurement of the economic harm suffered solely as a result of the changed site condition can be accomplished at a reasonable investment.

The process of assessing potential damages is a critical issue for all parties involved, not just for the team member preparing a claim for increased costs. Owners who were unaware that an environmental liability existed on their property are being faced with contaminated property that must be cleaned up before it can be used. An owner should have a good understanding of the types of costs that could arise from such exposure to evaluate the claims a contractor or others may submit. From the contractor or other team members' perspective, it is important to understand the documents that may be required to support a claim for damages resulting from a differing site condition. Those submitting a claim should also understand the various cost components to be considered in preparing and supporting such a claim.

§ 12.2 Existence of Differing Site Conditions

There are many situations in which the discovery of an environmental hazard on a construction site could be considered a differing site condition under traditional contract language. One situation could involve a contractor who discovers an environmental hazard on the site where a routine construction project is underway. For example, a contractor has begun construction on a multilevel parking garage and discovers drum tanks, some of which have corroded and are now leaking, buried in the soil where the foundation is to be erected. The contractor notifies the owner and may be under an obligation to also notify regulatory agencies.[1] The soil and groundwater may have to be tested, requiring the contractor to stop work while a remediation plan is developed. The contractor may or may not be qualified to perform the removal and proper disposal of the tanks. Meanwhile, the contractor may not be able to continue the construction as originally planned. Similarly, a contractor may be in the process of restoring or rehabilitating an older building. The contractor may have torn apart parts of the walls or ceiling, thereby releasing asbestos into the atmosphere. This discovery can also delay the renovations on the building as well as lead to increased costs.

Another situation involves a contractor who is under agreement to remediate the hazardous waste at a particular site. In this instance, the discovery of the hazardous waste is not a differing site condition. However, the discovery of quantities of hazardous waste that are different from the quantities specified in the contract and upon which the contractor based the project bid may constitute a claim of a differing site condition. Other situations that might be considered differing site conditions include the discovery of contaminants requiring different remediation techniques than those planned, the discovery of groundwater contamination, and the discovery of other environmental problems that go well beyond the remedial action initially planned for the site. Reconsideration by the governmental agency of the adequacy of the remedial action being undertaken

[1] 42 U.S.C. § 9603(a) (1988).

also can result in increased costs incurred by some or all of the project team members.

§ 12.3 Decision to Pursue Claim

Many factors are considered in making the decision to pursue a claim to recover increased costs. One of the more significant factors may be the process established for the parties to deal with changes during the project, as outlined in the contract and by the contracting procedures for the project. If this process maintains the balance of responsibility and risk between the owner and the other project members that was established through the original contracts, then the need for claims may be substantially diminished. For example, if the contractor and other project members immediately notify the owner when a differing site condition is encountered, the owner has the opportunity to evaluate the impact of the changed condition and discuss potential solutions and options with the project team. The parties may negotiate the appropriate change orders to the contracts to compensate the contractor and others for costs incurred due to the differing site condition while providing the owner with a means of control over the performance of the new scope of work.

Another significant factor may be the quality of the information that is available to support the claim. It is the filing party's responsibility to identify the costs, and only those costs that were incurred as a result of the differing site condition. Thus, to be viewed as credible, a claim should be supported by reasonable accounting and cost information. Ideally, these records allow for the segregation of costs associated with particular work activities. (As a practical matter, parties may want to consider establishing separate cost accounts at the start of the project to provide for the accounting of differing work activities.) If such segregation cannot be easily made or if the accounting records are not readily available, the time and effort needed to prepare a claim should be weighed against the benefit of successful resolution of the claim.

§ 12.4 Types of Claims

Claims to recover damages due to a differing site condition are generally made for compensable damages. Compensable damages can be divided into direct and consequential damages. Although these two types can be described generally, the determination of whether damages are direct or consequential is a matter of law.[2] *Direct damages* result from the increased costs associated with working through the differing site condition. If a contractor discovered an additional 2,000 cubic yards of contaminated soil to be excavated, all of the costs of

---

[2] Roanoke Hosp. Ass'n v. Doyle & Russell, Inc., 214 S.E.2d 155, 160 (Va. 1975).

removing the 2,000 cubic yards of contaminated soil could represent direct damages. Using that same example, assume that the additional time spent excavating the soil delayed the start of construction such that construction had to continue through the inclement weather season. The costs associated with weather delays could also represent direct damages.

Consequential damages are those damages that do not directly flow from the contract change but that may reasonably be expected to occur as a result of the change. The lost profits the owner suffered because the contractor was responsible for delays on the project is an example of consequential damages.

§ 12.5 Calculating Claims

The quantification or measurement of the increase in costs due to a differing site condition may be a complex process and may take differing forms depending on the facts and circumstances of each case as well as the parties involved. However, two approaches for calculating claims that are commonly followed are the *total cost approach* and the *cause-and-effect approach.*

In using the *total cost approach,* the increased costs associated with the changed condition are generally measured as the difference between the claimant's estimate or bid amount and the actual costs incurred in performing the work. This method of damage calculation is relatively simple and cost effective to prepare. However, the total cost method faces significant challenges when assessed by the receiving party. The total cost approach presents a claim in a manner that assumes that all amounts incurred over the original contract price were the result of the changed condition and that the contract price was reasonably established. Additionally, the total cost approach assumes that all cost increases from the bid amount are the responsibility of someone other than the claimant. A construction project is rarely completed entirely as planned or expected by any party. A claimant's increased costs could be the result of bid errors, cost reporting errors, and any number of other factors. Under the total cost approach, the claimant may be recovering costs (including profit and other charges) that are the responsibility of the claimant and to which the claimant is not entitled.

The *cause-and-effect approach* to calculating damages may reduce the number of challenges faced by the use of the total cost approach. The cause-and-effect approach establishes a causal link between the impacting event and the resulting cost increase or decrease. An examination of actual work performed and costs incurred associated with particular work activities is made. Costs can be examined chronologically to assess the impact of the differing site condition. Using this approach, cost increases can be assigned to differing site conditions, subcontractor delays or inefficiencies, changes in governing regulations and requirements, and other factors that can affect a project in an attempt to ensure that the claimant recovers only those price increases to which it is entitled.

The choice of a claim method or variation on a method can depend on the claimant's ability to produce accounting and cost records that can adequately segregate costs by work activity performed and the claimant's ability to isolate the impact of each event. Although a cause-and-effect approach may be a preferable method to use, there are instances in which the total cost approach may be the only practical method. The courts have established four conditions that must be met for the total cost method of calculating damages to be used.[3] These four conditions are:

1. There is no alternative method of calculating damages available
2. The original bid is reasonable
3. The actual costs incurred are reasonable
4. The claimant is not responsible for any of the additional costs incurred.

The choice of a method of claim preparation can be affected by contractual terms and possibly by certain liability theories known by the claimant's counsel. The liability theory or cause for making a claim and resulting support for a claim go hand in hand in almost every situation. With this in mind, the analysis performed to determine damages may also be used to support the claimant's liability position. For example, in calculating the labor component of a claim, it may be necessary to prepare a productivity analysis. The analysis that shows a loss of productivity also demonstrates the impact of the differing site condition on the labor efficiency of the work force. Thus, the calculation of damages associated with the loss of productivity can also be a tool in isolating the fact that the site condition had a direct impact on the claimant's performance.

In this regard, some contracts contain provisions that calculate estimated contract price based upon a unit price. Total contract price equals the unit price multiplied by the number of units of work contracted for (such as cubic yards of contaminated soil removed or drum tanks filled with toxic waste removed). If these provisions exist, a claim could be calculated based primarily upon unit prices. This does not mean that specific cost accounts should not be analyzed, but rather that the unit prices could be adjusted based on increased costs in the specific work components associated with the differing site conditions.

§ 12.6 Case Example

To illustrate some of these issues facing parties involved in a claim situation, a case example may be helpful. The case involves the cleanup of a site on the National Priorities List. The contractor agreed to furnish labor, materials, and equipment to carry out the cleanup at a fixed price. The cleanup primarily

[3] WRB Corp. v. United States, 183 Ct. Cl. 409 (1968).

involved on-site incineration as the remedial technique chosen. A number of changes took place that, the contractor alleged, caused it to incur additional costs—approximately double that of the contract price. In this instance, the contractor knew that the property contained hazardous waste but allegedly did not anticipate the type of waste or quantities of waste discovered.

Some of the events or changes at issue in this case relate directly to differing site conditions. For example, the contractor alleged that subsurface soils where the foundations were to be erected for the incinerators were discovered to be seriously contaminated to an elevation well below ground level. This discovery necessitated a change in anticipated excavation and construction methods (not included in the bid). The contractor also alleged that there were material discrepancies between the quantities of hazardous materials specified in the pre-bid documents and the actual quantities of contaminants found on site. These and other issues allegedly resulted in incineration costs far exceeding the price of the fixed-price contract.

§ 12.7 Evaluating Contractor's Bid

As mentioned in **§ 12.5**, in some instances the total cost approach may be the only method available for calculating damages. In those instances, the focus of an assessment of the claim is on the bid and estimates used as the basis from which to measure the claim as well as on the costs reported in the claim. Examining the claimant's bid and modifying the bid for errors and omissions that are detectable upon review can be an important step in assessing claims prepared under this approach. Errors and omissions commonly include clerical or judgmental errors involving, for example, identification of the construction location or the quantity of material required. In addition, it is important to inspect proposals and bids for work that is unique to a project.

As an example in an environmental remediation, the contractor's bid should be examined to determine if costs associated with performing work specifically called for by such a project had been adequately included in the bid or proposal (or has been inappropriately included in the claimed cost increase). The establishment of emergency procedures to minimize the impact of a problem arising during a hazardous waste remediation project may be the responsibility of the claimant party. Establishing these procedures is a necessary step to complete before the start of any remediation project. Similarly, the continued training of personnel in the proper handling of hazardous material as well as identifying and preparing for the site-specific nuances or procedures for remediation are all necessary activities. Also, there are extensive permitting requirements that may be the responsibility of the claimant and that can be both costly and time consuming. In conjunction with permitting requirements, there are requirements for the testing and documentation of the hazardous material. The actual testing and timing of the testing are critical considerations for which the contractor's reported costs should be evaluated against the contractor's bid. Costs associated with performing these procedures may have been necessary elements to be included in the contractor's bid. If they were not included, estimated costs to complete these efforts as well as any other identified bid omissions must be added to the bid (or reported costs removed from claimed costs) to determine a reasonable estimate of costs absent the differing site condition.

The contractor's bid and reported costs should also be evaluated to determine the impact on the claim of any change orders or other modifications that have been made to the contract. Modifications to the contract through change orders can have a significant impact on the overall value of the contract for claim computation purposes.

§ 12.8 Evaluating Contractor's Actual Costs

The calculation of actual reasonable costs can be a relatively straightforward process provided the contractor has maintained sufficiently detailed cost records. These cost records can then be used to support cost increases in many different cost accounts for the project.

Major cost accounts typically affected by a differing site condition include labor costs, equipment costs, disposal costs, overhead, and subcontractor costs. Other cost accounts that may be affected, but generally to a lesser extent, include material costs, financing or interest costs, and profit.

Labor costs. Labor is generally a significant component of overall cost on any construction project and on any environmental remediation project. If the claimant segregated employee hours associated with the work performed due to the differing site condition, the calculation of the labor component may be simply the additional hours worked multiplied by the appropriate wage rate. Sometimes the wage rate is stipulated in the contract. If not, the wage rate can be calculated to include payroll plus insurance and applicable taxes, applicable union fringe benefits, and the costs associated with supervision. However, if the contractor did not segregate costs, the calculation becomes more complicated.

A claimant may calculate increased labor costs based upon loss of productivity experienced on the project. Especially in a claim involving a differing site condition, a contractor may argue that the changed condition had a negative impact on the estimated productivity of the workers on which the bid was based. For projects involving environmental remediation, the type of hazardous materials being handled can have a significant impact on average productivity. Workers who are handling particular types of hazardous materials are required to wear varying levels of protective equipment. The more restrictive the required equipment is, the less efficient the worker may be. In addition, time spent removing and preparing such equipment can have a significant impact on worker productivity. One estimate of lost productivity based upon the protective equipment worn by workers was measured as follows:[4]

Normal Nontoxic Waste Removal	Level D Protection	Level C Protection	Level B Protection	Level A Protection
90–100%	85–95%	65–75%	40–60%	10–30%

Consequently, the calculation of lower productivity and the associated higher cost of labor can be a critical part of an increased cost claim.

In addition to lower productivity caused by the protective gear workers wear, workers who handle hazardous waste may be restricted on the number of hours they can work, which can also result in lower productivity.

In the case example in **§ 12.6**, the contractor claimed that the quantity of waste to be remediated varied significantly from the quantity stipulated in the contract. Additional workers might need to be added to the project team to maintain the overall contract schedule. Labor costs would consequently increase as a result of adding workers to the team. The mix of workers needed might also change, causing the average wage rate to increase.

Environmental remediation projects require the use of specialists who may not be part of the staff on a routine construction project. The cost of labor is increased by emergency medical technicians, health hygienists, lab technicians, and others on staff at remediation projects. The cost of these personnel needs to be considered and evaluated.

The personnel who work on environmental remediation projects undergo substantial medical testing before being allowed to work on these types of projects. Some workers may be prohibited for medical reasons from working on a remediation project that requires wearing a respirator. Additionally, the personnel receive substantial training in handling the hazardous material and in procedures to be followed in case of an emergency. Because of these requirements, it may be difficult for a contractor to add workers to the project or to move workers from one project to another project. In assessing damages on a routine construction project, one way a contractor may attempt to mitigate damages is through work resequencing. In the case of environmental remediation projects, there may be complicating factors requiring further analysis.

Equipment costs. The cost of equipment may also be a large component of the actual costs incurred on a project. Equipment may have sat idle due to unanticipated delays caused by the differing site condition or may have been on the site

[4] D'Annunzio & Renda, *Bidding a Hazardous Waste Project, in* Hazardous Waste Disposal and Underground Construction Law 283 (R. Cushman & B. Ficken eds., John Wiley & Sons 1987).

for a longer period of time because of added work. There are various techniques for calculating the cost of equipment. One technique computes the owner's cost of equipment by analyzing the recorded depreciation expense, maintenance expense, insurance expense, storage, tax, and other expenses incurred in operating and maintaining the equipment. These cost accounts are allocated to specific pieces of equipment to determine an hourly rate. This technique results in a cost rate for equipment that more closely measures the owner's cost of using the equipment.

Another common technique for calculating the cost of equipment relies upon published commercial rates to rent the piece of equipment if a rate has not been stipulated in the contract. These rates are published in industry guides such as the *Contractor's Equipment Manual* and the *Rental Rate Blue Book for Construction Equipment.* Additionally, it is important to note that there is a distinction between the rates for equipment that is being operated and equipment that is idle. The calculation of additional equipment costs may be the additional length of usage multiplied by the applicable cost rate. If costs have not been segregated so that additional usage can readily be determined, a correlation can be made to hours worked.

Environmental remediation projects add another dimension to calculating the cost of equipment. A remediation site may be divided into different zones. For instance, there may be a contaminated zone and a safe zone. Contaminated equipment cannot be moved into a safe zone without first being decontaminated. This can be both a costly and time-consuming process. The cost associated with decontaminating a piece of equipment and moving it to another zone where it can be used must be weighed against renting another piece of equipment while the contaminated piece of equipment sits idle.

Disposal costs. The costs of disposing of the hazardous material is another component of actual costs incurred. The disposal cost of waste is not typically significant in a routine construction project. However, disposal costs can be significant in a remediation project and greatly depend upon the type of waste involved and the remedial techniques chosen. Some contaminants can be stored, capped, or have some other on-site remedial action performed that could cost substantially less than off-site remediation options. Other contaminants require incineration to be disposed of in accordance with regulatory requirements, a technique that has a substantial cost whether on-site or off-site incineration is performed.

Minimizing disposal or remediation costs can begin by ensuring that a qualified and experienced remediation contractor is engaged to perform the necessary cleanup. On a number of sites where remediation of a hazardous waste was not anticipated, the on-site excavation or civil contractor was hired to perform the required remedial action. Increased costs and added project delays occurred in these situations as the contractor did not have the requisite regulatory reporting or remediation experience, resulting in an inefficient and ineffective cleanup.

The method and cost of transporting hazardous waste as well as labeling and testing waste disposed of in accordance with regulatory requirements can increase overall project costs. Assessing the change in these costs as a result of a particular event or circumstance may be particularly relevant in a differing site condition situation.

Overhead. A construction project or environmental remediation project typically incurs two types of overhead. The project receives an allocation of *home office overhead* as well as *field overhead.* Accepted formulas are available to allocate home office overhead. However, the formulas are usually used if there is not a way of determining overhead costs directly associated with the project.

Additionally, the contract may contain provisions that restrict the costs to be included in the overhead pools charged to the project. If so, those costs cannot be included in the allocation rate determined for claim purposes. An environmental remediation project requires extensive testing, monitoring, and inspection. The claimant may have the responsibility to manage and maintain the records associated with these administrative procedures. The costs are added to the routine field overhead accounts along with costs of supervision, field location administration, and quality control.

Other factors. There are other factors that should be considered in preparing or reviewing a claim for damages. For instance, architect/engineering fees for construction project redesign and drawing refinements may be necessary. If the site is to be abandoned or if structure redesign is necessary (avoiding the waste area), significant costs may be expended completing such activities.

In addition, if subcontractors are involved on the project, a contractor can either include the costs associated with the settlement of a subcontractor's claim in the contractor's damages claim or the subcontractor's claim can be passed through to the owner. If the subcontractor's claim is treated as a pass-through claim, it should be prepared in the same manner as the general contractor's claim.

Finally, the profit rate applied to such claims should be carefully prepared and scrutinized. Contract profit rates may or may not approximate the reported profit rate on a contract. Determining the appropriate rate through the review of performance may be the best method.

§ 12.9 Recovery of Environmental Damages

Some contracts contain provisions that prescribe procedures for the resolution of claims. Often a claimant can negotiate a settlement that is satisfactory to all parties through this process. If not, a claimant must seek remedy beyond this level. The claimant can appeal to state review boards or administrative boards.

Outside of litigation, disputes are also resolved through alternative dispute forums including arbitration, mediation, and minitrials.

An environmental remediation project or construction project can last many months or years. Throughout this time, an owner's financial position can change dramatically from its position when the contract was signed. It is difficult for a claimant to recover costs from another bankrupt party. In those instances, the claimant may attempt to recover from the owner's parent corporation or others if successor organizations have financial viability.

§ 12.10 —Recovery from Other Potentially Responsible Parties

The owner has several avenues to pursue to recover costs incurred for environmental cleanups and cost overruns on those cleanups. One alternative is for the owner to recover costs from other potentially responsible parties (PRPs). This, of course, depends on the actual facts and circumstances surrounding the owner's environmental situation. For example, an owner may have the ability to recover damages due to environmental claims from various other parties who may have been responsible for the original disposal or transportation of waste to the site or the operation of an environmental site. Assume an owner purchased a piece of property and now is constructing a multi-unit apartment complex. The contractor, upon excavating soil, discovers several underground storage tanks as well as several hundred buried 55-gallon steel drums. The contractor, of course, will look to the owner for either recovery of the increased costs for removing the hazardous substances or, more likely, will look to the owner to hire a remediation contractor. The owner, in turn, may look to other PRPs who previously operated and sent waste to the site.

PRPs or "covered persons" are generally defined under the Comprehensive Environmental Response, Compensation and Liability Act (CERCLA)[5] as:

1. The current owner or operator of the facility from which the hazardous waste emanates or threatens to do so
2. The owner or operator of a facility at the time of the disposal of a hazardous waste at the facility
3. Any person who arranges or arranged for disposal or treatment of hazardous wastes
4. Any person who accepts or accepted hazardous waste for transport to a facility.

The owner's success at identifying these parties depends on several factors, including the documentation available from prior periods. The owner can hire a

[5] 42 U.S.C. § 9607(a) (1988).

professional organization skilled in performing such investigations and other analyses of records to assist in the process. If and when such other parties can be identified and information about their role and activities in generating the waste, transporting the waste, or operating the site is obtained, then the task arises as to how to recover the cleanup costs.

Liability under CERCLA is strict, joint, and several. Essentially, any single PRP may be held responsible for the cost of the cleanup of the entire site. However, cleanup costs are typically allocated among PRPs. This process can be lengthy and costly, and the owner must perform a cost-benefit analysis as to the time required to recover such damages as well as the likelihood of such a recovery. Of course, the owner must weigh this against any delay that might be created (and resulting increased costs and lost profits associated with a delay) if an owner chooses to have additional parties participate in the remediation decision-making process. Depending on the complexity of the remediation action and the estimates of costs the owner receives from reliable remediation firms, the owner may decide to perform the cleanup, fund the costs, and pursue recovery at a later point in time. On the other hand, the owner may not have the ability to fund such costs and may have to pursue others to assist in the initial funding to begin the remediation process.

In substantiating the owner's claim in either case, the owner must obtain reliable cost estimates and actual cost information from the remediation design firms as well as the contractors. The owner's ability to demonstrate prudent management of the remediation process may certainly enhance an opportunity to negotiate participation by other PRPs in the remediation funding process. However, even if the contractor takes all the necessary precautions, long and protracted litigation may still need to be undertaken.

Assuming the owner hires a contractor to perform the remediation and then attempts to recover these costs from other PRPs, those other parties will most likely insist on a review and assessment of the costs and procedures used to remediate the site. These parties will scrutinize the remediation techniques and costs, and the owner will most likely be in the position of needing to prove that the decisions made in the remediation process, as far as the selection of the technique to be used and the selection of a remediation contractor, were efficient and cost effective. The Environmental Protection Agency (EPA) has established nine criteria against which potential remediation techniques can be measured. The two threshold criteria that must be met for the technique to be selected are:

1. Overall protection of human health and environment
2. Compliance with applicable or relevant and appropriate requirements.

Five other criteria are divided into balancing criteria:

3. Long-term effectiveness and permanence
4. Reduction of toxicity, mobility, and volume through treatment

5. Short-term effectiveness
6. Implementability
7. Cost.

The final two are modifying criteria:

8. State acceptance
9. Community acceptance.[6]

The owner can measure the remediation techniques available against these criteria and can respond to scrutiny from others based upon these criteria.

Therefore, the owner should be very careful in its selection of remediation contractors and its project management overseer. The owner may establish its own criteria in selecting a remediation contractor. These criteria may include a history of successful completion of similar remediation projects, sound financial condition, knowledge of changing remediation techniques and technology, as well as overall good project management skills required in any construction-type project. The owner may be able to measure the remediation contractor's experience based upon the quality of the bid submitted by the remediation contractor.

In allocating costs among PRPs, there are no consistent rules or guidelines to use in every situation. There are, of course, the factors enumerated in the amendments written by Representative Albert Gore, which courts have used in certain cases to assist in allocation of costs. The Gore amendments attempt to provide guidelines on apportioning liability based on the evidence contributing parties present in court. The following factors could specifically be considered:[7]

1. The ability of the parties to demonstrate that their contribution to a discharge, release, or disposal of a hazardous waste can be distinguished
2. The amount of the hazardous waste involved
3. The degree of toxicity of the hazardous waste involved
4. The degree of involvement by the parties in the generation, transportation, treatment, storage, or disposal of the hazardous waste
5. The degree of care exercised by the parties with respect to the hazardous waste concerned, taking into account the characteristics of such hazardous waste
6. The degree of cooperation by the parties with federal, state, or local officials to prevent any harm to the public health or the environment.

Most cost allocation procedures, however, have been performed in private settings outside of court by groups of PRPs through use of independent

---

[6] 40 C.F.R. pt. 300 (1991).

[7] 126 Cong. Rec. H9461 (daily ed. Sept. 23, 1980).

consultants. The independent consultants bring objectivity to the analysis as well as the financial and economic expertise needed to develop a model that meets the parties' objectives.

Cleanup costs can be assigned to other PRPs based on varying cost allocation models. These models can allocate costs based upon volumetric contributions the parties made and on the duration of the contributing parties' use of the site. Allocation of costs can also be based on the toxicity of substances disposed of by the contributing parties. In addition, costs can be allocated to parties based on a model of cause and effect otherwise known as a *cost-driven approach.* Such an approach identifies the incremental cost of differing remedial techniques implemented at a site and assigns each cost increment to those parties responsible for the higher level of contamination. These processes are never easy and can be contentious as questions arise concerning who should pay for what and why. The relative liability among generators, transporters, prior owners, and current owners is not a science or an easy determination. It behooves the owner to hire an experienced environmental attorney and consultant to assist in negotiations and/or litigation to pursue these claims against PRPs.

§ 12.11 —Recovery from Owner's Insurance Carriers

An owner can also attempt to recover its costs or claims associated with the cleanup from its general liability insurance carrier. Whether a company has insurance coverage for environmental damages is a relatively complicated question today. Not only does it require an in-depth review of the types of environmental costs that are being incurred, but it also requires an in-depth legal analysis and possibly litigation related to a variety of insurance policy issues. Many companies purchased insurance over a number of years at primary and excess levels and are now looking to their insurance carriers for coverage due to environmental damage claims. To better understand the contested issues regarding insurance coverage for damages related to environmental exposure, it is necessary to recount some of the history of the germane provisions found in insurance policies.

The insurance industry over the years has written several types of general liability policies for policyholders. Pre-1966, there were generally policies that covered accidents and did not exclude environmental harm. After 1966, policies were occurrence based and included coverage for damage the insured neither expected or intended. During the early 1970s, pollution exclusion clauses were added to most comprehensive general liability (CGL) policies. The clauses typically stated that the policy does not apply:

> to bodily injury or property damage arising out of the discharge, dispersal, release or escape of smoke, vapors, soot, fumes, acids, alkalis, toxic chemicals, liquids or

gases, waste materials or other irritants, contaminants or pollutants into or upon land, the atmosphere or any water course or body of water; but this exclusion does not apply if such discharge, dispersal, release or escape is sudden and accidental.[8]

Claims arising from pollution, except for pollution that was the result of a "sudden and accidental" discharge, were not covered under these CGL policies.

Insurers, policyholders, and even the courts, however, have differing opinions on the meaning of "sudden and accidental." Policyholders argue that the "sudden and accidental" exception applies in any instance in which the policyholder did not intend to cause the resultant damage. Insurers believe that the discharge must be both *sudden,* meaning occurring over a short time, and the damage must be both *unexpected* and *unintended* to avoid triggering the pollution exclusion. The courts have not provided a consistent answer.

As a result of the extensive litigation generated by the conflicting interpretations of the exception, most CGL policies in the 1980s included absolute pollution exclusions without the exception for "sudden and accidental" discharges. Policyholders or owners who have policies written in the 1980s have little hope of recovering damages resulting from environmental cleanup efforts from their insurance companies. However, for policyholders or owners who had policies providing coverage written in the 1970s and earlier, the issue of recovering damages for environmental cleanup from their insurance companies is open to debate.

There are differences in opinion as to whether environmental cleanup costs are damages covered under the language of an insurance policy. Many insurers have argued that cleanup costs are equitable relief and are consequently not covered.

Timing is another critical issue in determining whether an owner can recover damages resulting from environmental exposure. The timing issue, often called the *trigger* of coverage, relates to what must occur during the period covered by the policy for the policyholder to receive coverage from the insurance carrier. Consequently, even if the discharge of waste meets the definition of property damage, there may be a question of which policy will respond based on interpretation of the coverage and when the trigger begins and ends.

Many knowledgeable attorneys believe there is not yet a precedent set as to whether insurance policies will cover environmental damage claims. Of course, policyholders believe that in the end, they will obtain coverage. On the other hand, insurance carriers believe they will avoid or minimize their payments of environmental damages. Some believe that the environmental problems are so large that neither corporate America nor the insurance industry alone can fund environmental cleanup, and therefore a combination of corporate funding,

---

[8] Miller & Lafebvre, Miller's Standard Insurance Policies Annotated 411 (1986) (quoted by Rutkin, *Absolute Pollution Exclusions: Absolute Bars to Pollution Claims,* Envtl. Claims J., Summer 1991, at 401–10).

insurance funding, and tax funding is needed to pay for the estimated billions and billions of dollars of environmental cleanup.

In any case, an owner should give notice to its insurers that it has a claim for environmental damages. An owner should retain counsel to determine whether potential coverage exists and to pursue the various avenues for either negotiation or litigation with the carriers. Here again, however, the owner should weigh the cost benefit of pursuing recovery from insurance because it will most likely be a very expensive, time-consuming process. The owner must analyze the financial impact it will have on the overall project from a delay disruption and increased costs perspective. The bottom line decision will generally be whether the owner can fund such cleanup and then pursue recovery from its insurers, or whether the project must be stalled until the insurance issue is resolved.

§ 12.12 Other Case Examples

Following are some examples of recent cases in the environmental insurance coverage area. This section by no means attempts to cover all relevant cases or all major issues, but provides an overview of the cases, the issues, and the decisions that may affect ongoing cases. There have been many decisions by various courts in the country, many in favor of policyholders and many in favor of insurers. These case examples indicate how various recent decisions may help or hinder the ability for an owner to recover environmental damages.

Example 1: Assume an owner bought a piece of property from a municipality. The property was purchased 15 years ago. The owner is in the process of building a high rise. In the initial phases of construction, it is discovered that, unbeknownst to the owner, the municipality had used the site for overflow disposal of municipal waste 50 years ago. In most cases, municipalities have not been held liable for environmental cleanup. Recently, however, a Second Circuit decision[9] stated that municipalities can be held liable for cleanup of sites at which they disposed. Therefore, although certain courts will state that the municipality is not responsible for cleanup, the owner may have the ability to recover from the municipality as a PRP.

Example 2: A company in California has owned property for 30 years. The owner provided underground storage services for various waste generators in several tanks on site. The storage of the waste occurred from 1962 through 1970. It is discovered in 1992, as the owner is in the process of removing the storage tanks to construct a shopping mall, that the underground storage tanks have been leaking since the late 1960s. There has been a recent ruling[10] in the Montrose

[9] B.F. Goodrich Co. v. Murtha, 958 F.2d 1192 (2d Cir. 1992).

[10] Montrose Chem. Corp. v. Canadian Universal Ins. Co., No. C594148 (Cal. 1990).

Chemical case in California's Second District Court of Appeals that a policyholder can now tap all CGL policies from the time that the pollution begins until the time the liability is known. Thus, this case provides for a continuous trigger of insurance policies. This owner could attempt to recover its environmental damages from all of its insurers that wrote coverage from the late 1960s through the current date.

Example 3: If an owner participates in environmental cleanup where there is not a court order or a mandate by the EPA, questions arise in the insurance context as to whether the policies will pay because the damages are voluntary payments. There can be certain exclusions for voluntary payments under the comprehensive general liability policies. The Augat, Inc. case[11] in Massachusetts is an example in which payments voluntarily made fell within the meaning of an exclusion and were therefore not covered under the policy. Another case, Intel Corporation,[12] in the Ninth Circuit Court of Appeals, held that when a policyholder enters into a consent decree with the EPA to voluntarily clean up polluted property, the costs are covered as damages under the terms of the CGL policy and are not subject to the pollution exclusion. In any case, owners should be careful in how they approach the payment and recovery of damages when they are involved in a potentially voluntary situation rather than a situation mandated by a court, the EPA, or a state agency.

Example 4: Assume an owner discovers that underground storage tanks on its property have been leaking contaminants into the surrounding soil for a period of years. In cases with similar circumstances, the courts have been divided over the interpretation of the discharge as sudden and accidental. The California Superior Court held in the ACL Technologies, Inc.[13] case that the term *sudden* refers to something that happens abruptly. Because the discharge occurred over a period of years, the pollution exclusion was applied. Similarly, the Michigan Supreme Court held in the Upjohn Company[14] case that the term *sudden* had a temporal element. The pollution exclusion was applied. In contrast, the Supreme Court of New York held in the Halstead Oil Company[15] case that the discharge was not readily discoverable by Halstead Oil and, consequently, was sudden and accidental. The pollution exclusion was not applied. Therefore, although certain courts state that discharge over a period of years is

---

[11] Augat, Inc. v. Liberty Mut. Ins. Co., 571 N.E.2d 357 (Mass. 1991).

[12] Intel Corp. v. Hartford Accident & Indem. Co., 952 F.2d 1551 (9th Cir. 1992).

[13] ACL Technologies, Inc. v. Northbrook Property & Casualty Ins. Co., No. X-61-95-76 (Cal. App. Dep't Super Ct. 1991).

[14] Upjohn Co. v. New Hampshire Ins. Co., Nos. 86906-86908 (Mich. 1991).

[15] Halstead Oil Co. v. Northern Ins. Co., No. 7043/90 (N.Y. Sup. Ct. 1990).

not "sudden and accidental," an owner may be able to recover from its insurance carriers in similar situations.

Example 5: Assume that the EPA orders an owner to perform a cleanup of the contamination on its property. The owner seeks coverage by its insurer for the costs incurred in responding to the EPA. Recently, the Washington Supreme Court held in the Boeing[16] case that response costs incurred by Boeing constitute damages covered under CGL policies issued by the insurers and are not subject to the pollution exclusion.

---

[16] Boeing Co. v. Aetna Casualty & Sur. Co., 113 Wash. 2d 869, 784 P.2d 507 (1990).

CHAPTER 13

LITIGATION PERTAINING TO ASBESTOS CONTRACTING

John E. Osborn, Esquire

John E. Osborn practices law in the Manhattan law firm of John E. Osborn, P.C., concentrating in environmental law and construction and real estate litigation. His practice involves complex construction litigation and comprehensive counseling on construction and environmental issues, including development of long-term strategies for risk management and litigation cost containment through use of alternative dispute resolution. A member of the faculty of New York University, Real Estate Institute and the Pratt Institute School of Architecture, Mr. Osborn is the author of several articles relating to asbestos contracts, operations and maintenance programs, risk management, and litigation. He is certified by the United States Environmental Protection Agency and the New York City Department of Environmental Protection to teach asbestos licensing courses.

§ 13.1 Asbestos Abatement Examples

Strategies set out by owners and property managers most often determine whether and to what extent there will be a contest over differing site conditions.

Even before a minute of analysis goes into the planning of a project, the budget and the experience of the owner, together with the time schedule within which the asbestos abatement work must be completed, will help foretell whether the owner is likely to face a claim for differing site conditions. If the owner wants the job done cheaply, is in a rush, and is ignorant of the risks involved in asbestos abatement, there is a very good chance that a pattern leading to a differing site conditions claim will be set early on.

Following are three examples that illustrate changed conditions or differing site conditions claims that could have been avoided or limited. The remainder of the chapter provides an agenda for avoiding these expensive pitfalls.

Example 1

A 12-story office building is about to be demolished so that the owner can construct a new building on the property. A great deal of attention has been placed upon the new structure, and a schedule for demolition, construction, and occupancy has been set.

Now the focus is on demolition. Under local governmental requirements, the asbestos in the building must be removed before demolition. The owner hires an environmental consultant who will design the asbestos abatement, make an award to the asbestos abatement contractor, and oversee contract administration for the abatement.

The owner urges the environmental consultant to undertake the project on a *turnkey* basis, in which the consultant takes full responsibility for the completion of the abatement, from design through final sign-off.

The compensation for the entire project is to be included in the bid the consultant places along with the contractor. The owner expects the contractor to be a subcontractor to the consultant.

The consultant refuses to enter an agreement to act as a turnkey contractor for the entire project and enters a contract with the owner under which the consultant will design and administer the abatement project. The consultant also agrees to assist the owner in choosing an abatement contractor.

The consultant advises the owner that a mere visual inspection of the premises, which might be adequate in a renovation project, is not good enough here, as every bit of asbestos (which could be left alone if this were a renovation) will be found as the building is being torn down. Clearly, the cost and delay will be greater if the asbestos is found during the job rather than pre-bid.

The consultant explains that destructive testing must be undertaken. (Destructive testing is the process of spot checking by taking apart the building components to determine conditions present. It can be done in preparation for demolition, renovation, or construction.) Reluctantly, the owner agrees to the additional expense for this testing, but only to the extent of examining "typical conditions" on each floor. A design is developed based upon the typical conditions, and the work is opened for bids.

The bid is awarded to an asbestos abatement contractor on a lump-sum price, and the contractor begins work. As soon as the abatement is begun, it is clear that in the building, which had been occupied by many different types of business tenants over the years, had no such thing as "typical conditions." On some floors, the tenants had removed the vinyl asbestos floor tile and replaced it with carpeting; on other floors, the vinyl asbestos tile had been covered with carpet. On some floors, the column covers had been taken apart and asbestos pipe covering had been removed, while on other floors the asbestos pipe covering remained. On some floors, no asbestos ceiling tile was evident, as it had been covered by other ceiling tile without asbestos, while on other floors, the asbestos-containing ceiling tile had been removed before the new tile was installed.

As it is discovered that, in each instance, conditions are not as "typically" shown on the contract drawings, the project is stopped, and a meeting is called to discuss how much additional compensation the contractor is to receive for abating the asbestos it has located that is beyond what was anticipated.

The contract provided that the contractor was entitled to be paid additional amounts if changed conditions were encountered. Ultimately, extra compensation was agreed to, and the asbestos abatement was completed. Although the price paid to the contractor was reasonable, the owner did not have the advantage of the bargain prices it would have received as a part of a lump-sum bid.

How the problem could have been avoided. The owner could have derived a better bargain had it chosen to investigate the premises further. This would have required further payment to the environmental consultant, probably including destructive testing on each floor. This way, the contract documents would have been complete, and the contractor's original price would have been complete.

Another solution owners sometimes choose is to include a clause making the contractor responsible for changed conditions when encountered.

This approach is always a problem, as it will cause the contractor to build in a contingency to cover unknowns. If the changed condition is significant enough, the owner may have no choice but to make additional payment to the contractor to avoid pushing the contractor into default or bankruptcy.

Example 2

A company is about to refinance its corporate headquarters, move out of New York, and then rent out the office space in that headquarters. Before the refinancing can take place, the lender for the company requires that a Phase I site assessment (a preliminary survey of site conditions that does not include probing into building components) take place and that any asbestos in fair or poor condition be removed.

The company's finance director negotiates the agreement with the environmental consultant to perform the Phase I assessment, and provision for destructive testing on floors 2, 11, 12, 22, 34, and 37 is made. The areas for destructive testing are set by the finance director, who indicates that they should provide a "representative sample" of the building conditions (but in fact they were chosen to keep the consultant from finding asbestos).

Once the visual inspection is complete, the environmental consultant writes a report that is used as a basis for obtaining a bid from an asbestos abatement contractor. The study, which becomes a part of the bid documents in its entirety, shows that it is only in certain portions of the structure that spray-on fireproofing is found in the building, and that there is no asbestos pipe covering on the heating piping that serves the building.

A contractor is retained based upon a lump-sum price to "remove all asbestos" in the building. The bid is based upon the findings of the consultant rather than on a full set of plans and specifications.

Once work is begun on the spot removals indicated to be necessary in the report, it becomes clear that there is spray-on fireproofing throughout the building and that the heating pipes, except on certain recently renovated floors, have asbestos-containing pipe covering.

The contractor and owner meet to resolve how to proceed. Although the negotiations progress in a reasonable fashion, the price is no bargain for the owner.

How the problem could have been avoided. Once again, a greater investment in the efforts of the environmental consultant would have allowed the owner to obtain a more accurate bid. In both cases, had the owner kept better records on the 'as-built' conditions of the building and disclosed these records to the consultant, there would have been no changed condition claim.

Example 3

A financial institution is eager to set up a center that will serve as a flagship for marketing a new financial product. An elaborate retail storefront operation will be built in a prime location to house this effort.

Although the institution already occupies the space to be used for the new center, regulations governing asbestos were instituted since the date of the last renovation. A filing must be made before further renovation is undertaken, certifying that the asbestos will be removed as part of the work.

Speed is critical in this process: the center must be opened within three months to confide with advertising during the tax season, when the product will be sold. To further complicate matters, the company does not want to close the site's retail use until a week before the asbestos abatement is to take place. This is significant because, although the asbestos consultant has access during off hours for visual inspection, there is no opportunity for the abatement contractor to do destructive testing before the bid. A lump-sum agreement is preferred because management has a set budget for the entire renovation. Although the contractor's bid must come in well before the existing use of the space ceases and the contract calls for the contractor to be fully familiar with the premises and to "visit the site" beforehand, this clause is not meaningful because the contractor is not provided access at all during the bid period. Also note that the retail space is located in an arcade/terminal structure that is below grade at one end of the retail space.

The lump-sum bid comes in from the four contractors solicited. The bids range from a low bid of $800,000 to a bid of $1,250,000. The project is commenced by the low bidder on a fast-track schedule and is just as abruptly stopped when (1) ducts above the space are found to be covered with asbestos; (2) a wall built over 100 years ago has holes in it that allow air to escape to the street (the wall is also covered by asbestos); and (3) trestles that compose part of the below-grade portion of the premises are partially coated, apparently inadvertently, with spray-on fireproofing and must be abated (which will be a slow and expensive process).

Negotiations take place to determine the pricing for the newly discovered components of the project. Once again the contractor and owner resolve the issue and the project is kept on target.

How the problem could have been avoided. Further owner expenditures on the environmental consultant investigation and better owner research of its own records would have saved a significant amount of expense. Hasty planning required to keep a project on schedule and acceleration of the work are costly items.

§ 13.2 Defining the Claim: Responsibility and Liability

A developer once said, "Building a high rise, once you have done one, is not a challenge; you tell me how many floors, and I can tell you how much it should cost. It is really just 'cookie cutter' work."

This is certainly not the case in renovation work. As urban areas have embarked upon restoring office buildings, housing stock, zoos, airports, skating rinks, train stations, and brownstones, the cost of doing so has been almost impossible to determine. One cannot know the result of a few hundred years of use, interim renovation, weathering, and decay unless the building is torn down and thoroughly checked.

Even the seasoned architect or engineer has learned not to speculate about the condition of the structural beams or the steel girders that have not been seen by anyone still living. The plans that show what is inside or behind a wall also may be outdated or inaccurate, if they can be found at all.

Issues of responsibility and liability make estimating the asbestos project even more of a problem. Among the architect, the owner, the general contractor, the environmental consultant, and the demolition contractor, no one wants to admit responsibility for overseeing the asbestos abatement. In some cases, regulatory constraints that must be met in connection with abating asbestos may dictate who will take on overall responsibility for the task. However, in other instances no one person is clearly in charge of asbestos abatement. This can present the biggest problem for all concerned.

This reluctance to take on responsibility for asbestos abatement is fairly recent. In the mid 1980s, there was an initial surge of interest in becoming an asbestos abatement contractor because the profit margin was seen to be substantial. As the construction industry has become more familiar with asbestos management, profit margins have narrowed, and the willingness to take risks has dwindled.

For the architect, errors and omissions (malpractice) insurance specifically excludes environmental risks. The design professional or environmental consultant who is eligible for errors and omissions insurance for the asbestos project can purchase it, although it is expensive.

Generally, in the beginning, when it came to asbestos abatement, the major participants in the construction process—design professionals, construction managers, and contractors—lacked expertise and insurance. Although risk management experts would certainly have advised against it, some entrepreneurs nevertheless were willing to perform asbestos abatement in exchange for the promise of a hefty profit margin.

Overall there is a lack of standards for those who design, perform, and monitor asbestos abatement projects.

Environment consultant. Other than training courses and certifications required by the federal, state, and local governments, the threshold knowledge, training, and experience of those who design and monitor abatement projects varies widely. The environmental consultant may range from the very inexperienced hygienist to the seasoned engineer.

Although it may be possible to tell the difference, even seasoned owners will have trouble distinguishing among different levels of experience, education credentials, and price. As discussed in **§ 13.7**, it is clear that a lack of standards leads to unevenness when it comes to estimating the asbestos abatement project or even developing a consistent understanding of what is involved in the project.

Asbestos abatement contractor. There is simply no standardization of credentials that helps the owner choose an asbestos abatement contractor. Often the owner will simply advertise for bids and award the project to the low bidder.

Due to health and safety concerns unique to environmental work, it is imperative that careful underwriting be done before an asbestos abatement contractor is chosen. Although at one time it was difficult to ascertain the track record or stability of a company, this should no longer be the case.

There are many distinctions that go beyond the basics. The basis for comparing abatement contractors may be their dissimilarities. Some began as insulation contractors, demolition contractors, general contractors, or drywall or carpentry contractors. Some started with absolutely no background in construction.

A complicated renovation project involving abatement requires an abatement contractor with an in-depth background in demolition and drywall work. In an abatement project involving removal of asbestos pipe covering which is to be followed by recovering, an abatement contractor with experience in insulation work may be best. Certainly, the estimate should be more reliable if the abatement contractor is aware of the next step. It may also serve the owner to award the drywall work to the same contractor.

Owner. The disparity in environmental consultants and contractors is magnified by the disparity among those owners who hire them. Of course, there is no conceivable guideline for ownership of property. There are those owners who will operate with absolutely no regard for health, safety, and environmental matters because of ignorance, concern for the bottom line, or other reasons.

Developing a list of credentials including education, industry experience, representative clients, and violations (or lack thereof) for each of the three entities discussed—owner, environmental consultant, and abatement contractor—should help determine whether a project will face a differing site conditions claim.

The profile of the owner may be the most enlightening and the hardest one to compile, as the owner is not bidding on projects or marketing its abilities to manage a facility. Following is a list of relevant questions to answer about the owner and its premises:

1. Are there accurate and up-to-date as-built drawings of the premises?
2. What are the qualifications, experience, and training of the managers of the facility? What is the level of staffing?
3. Is the facility to be abated in good physical condition as to long-term maintenance? Is it clear of debris? Are the elevators in good working order? Are there staging areas for abatement work? Are there elevators that will be available for use during the project? What is the condition of the electrical system? Is there an ability to turn off or disconnect the electricity during asbestos abatement work?
4. Is the owner or are the facilities managers forthcoming in answering questions? Are they responsive and cooperative on the walk-through?

With built-in complications inherent in the asbestos abatement process and a wide disparity among the participants, it is no wonder that there are few asbestos abatement projects without a basis for some differing site conditions claim.

§ 13.3 Factors Leading to Claims

The prevalence of subjective factors in asbestos abatement work leads to differing site conditions claims. Determining what the experienced, prudent contractor would have known is hard to judge.

Probably the worst case scenario is one in which the owner is unsophisticated, is looking to cut corners, and therefore hires only a minimal amount of consultant assistance and aims to retain a "cheap" contractor. In this instance, it is an open issue as to what constitutes a changed condition. The contractor walked onto the job without a clear delineation of what the conditions were at the premises because the consultant was only paid a minimal amount to draw up the abatement plan and to allow the owner to merely comply with regulatory requirements. Subject to contract provisions and the provable state of knowledge of each of the parties, virtually every unknown in this case turns into a differing site conditions claim for the contractor.

Differing owner knowledge is critical and may be as follows:

1. The owner knows and the contractor does not know of the condition: when the condition is encountered, the owner must pay the contractor, even in the face of an exculpatory clause.
2. The owner does not know and the contractor does not know of the condition: when the condition is encountered, the owner will have to pay unless there is an exculpatory clause that provides otherwise.
3. Variations of these circumstances occur when the owner should have known or the contractor should have known.

Although differing site conditions claims in asbestos abatement are most closely associated with changes in perceived conditions at the site, changed conditions can relate to factors other than those immediately at hand.

Regulatory environment changes. As environmental regulations change rapidly, the cost of performance of asbestos abatement work can be dramatically altered midstream. The vigor or the prevalence of enforcement of existing laws and regulations can also change, increasing the cost of performance. An example is a regulatory requirement that a contractor use transfer stations for the asbestos waste, assuming that the requirement was not being enforced at the time of the contract signing but is being enforced at the time of performance. If the fee of the transfer station was not included in the bid, it is not clear whether the contractor will be able to get the owner to pay. A careful review of contact clauses and the knowledge of each party will have to be made.

Changes in availability of waste disposal sites. Localities rebelling against taking disposal of environmental waste and limited capacity of disposal sites may lead to a dramatic and unexpected increase in cost to the contractor for disposal. This problem is probably best dealt with by providing, from the outset, that the owner has full responsibility for liability and cost. This eliminates the contingency that the abatement contractor will build into the bid to cover the risk of liability or unexpected cost increases attendant to disposal of asbestos waste.

Changes in labor arrangements. Labor relations can always cause problems in the construction industry, and asbestos abatement projects are no different. There is the perception that because the trades performing abatement work operate under the cover of darkness (often literally at night or on weekends or inside containment areas) and they often are on and off the project before following trades arrive, they can use nonunion labor on an otherwise union project.

If this is discovered, there will often be pressure on the owner to make sure that the abatement contractor uses union labor. Most often the standard construction contract includes a clause making the abatement contractor responsible for harmony of labor relations, and the contractor will be forced to pay additional increments due for union labor.

Although owners have been known to be sympathetic to the plight of the contractor in this circumstance, concessions will most often be voluntary and only partial. Other owners may justifiably believe that the contractor's assumptions in estimating labor for a lump-sum bid are not the owner's concern. Although there are contractors who disagree, this is not a true differing site conditions claim.

Negative air patent restrictions. Those in the asbestos abatement industry must be aware that GPAC, Inc. licenses contractors to use the patented negative air system. The patent provides a broad scope of protection for asbestos abatement using the application of negative pressure and filtration of air by the use of negative air machines. Since its issuance in 1986, the patent has repeatedly been challenged and issues contesting the validity of the patent are currently pending before federal district courts in North Carolina, Delaware, and Washington, D.C. In an effort to enforce its patent, GPAC, Inc. has embarked on a campaign to force contractors to obtain a license before employing the system. It has also notified building owners and even consultants about the existence of the patent. In order to protect themselves, Mississippi and New Jersey have implemented policies requiring that bidders either hold a license from GPAC or agree to indemnify the public entity from any potential infringement claims.

Until the courts determine the validity of the patent, contractors, builders, and consultants must remain wary. In order to minimize potential liability from patent infringement claims, legal counsel should be sought in drafting bids and contracts involving asbestos abatement.

Patent requirements. With the issue of the GPAC patent having been contested in the courts, its importance cannot be minimized. If the contract is signed without any regulatory or contractual responsibility for use of patented equipment, will the contractor prevail in its attempt to recover for the cost of complying with the license fees to use the patented equipment if the holder of the patent requires it after the date the contract is signed? Is the answer affected if the patented equipment must be used to comply with regulatory requirements? Is the answer affected if the patented equipment is not required by regulators but the use of any other acceptable method will cost the contractor double the money?

It seems appropriate that if the design calls for a proprietary item to be used and it has not been disclosed that the item is patented, the contractor should not be required to bear the cost of defending a lawsuit and paying a settlement or judgment in relation to an infringement suit or for additional license or royalty fees. However, if the contractor has utilized a patented item on its own, then the contractor might be expected to bear that cost.

The contractor should indemnify and hold harmless and pay the cost of defense of the owner and environmental consultant if they are sued for patent infringement when the contractor has chosen the patented equipment.[1]

Owner interference. The asbestos abatement project is especially susceptible to disruption and massive additional cost if the owner does not fully cooperate and "clear the way" for the contractor's performance. The failure of these obligations may be considered a changed condition entitling the contractor to be paid more. Some of the owner's shortcomings in this area include:

1. **Failure to provide access.** This may be particularly costly even in the short term, and the owner's responsibilities may be significant. For example, the owner may have been required to provide contractor access to tenant space by providing notice to tenants to vacate their space during a weekend or after-hour period when the contractor would be working. If the contractor shows up but the owner forgot to notify tenants and the contractor is not able to work that night or that weekend, the cost could be very considerable.
2. **Decision making of the owner/consultant.** If the decision-making process of the owner or its consultant is slow and holds up the contractor's performance, the contractor could be entitled to compensation. This is particularly the case when the contractor is working overnight or over a weekend and is unable to reach a representative who is authorized to make a decision.
3. **Use of elevators.** If the contractor's bid is predicated upon using the elevators and the stairways must be used to carry out debris, this may entitle the contractor to be paid more. This issue should be addressed pre-bid.
4. **Water supply.** If the contractor is promised in the contract that it will be allowed to use water supplied by the owner and then is required to run temporary water lines, the contractor may be entitled to additional payment.

[1] J. Osborn, The National Insulation and Abatement Contractors Association—NIAC Owner-Contractor Asbestos Abatement Agreement Guide § 1 (2d ed. 1990). This guide may be obtained from NIAC at 99 Canal Center Plaza, Suite 222, Alexandria, Va. 22314.

5. **Electricity.** In a fairly common problem in renovation and abatement, before the abatement is undertaken, it is assumed that the building power can be used for the contractor's equipment. If once the job is begun the power must be shut off, the contractor may be entitled to compensation for supplying temporary power and power lines. The abatement contractor may be the one to discover this, as it is the first one with access to the site. This must be addressed in the contract.
6. **Use of staging areas.** Of paramount importance on the abatement project is where the asbestos that is being removed can be placed. If there is no temporary staging area, cost implications could be severe.
7. **Failure to provide information.** When the bid documents provide that the contractor will be given "as builts" of the area being abated, the failure of the owner to do so may later bring about a claim by the contractor for additional compensation. It would appear that the failure to provide consultant reports when they have been promised may also lead to a claim for changed conditions.

§ 13.4 Contract Clauses: How They Can Help or Hurt

The quality of the contract documents, from the boilerplate to the significant business terms to the technical specifications, really sets the pattern for the project. The abatement project, especially a costly one done under tight time constraints, demonstrates the huge dollar amount attributable to each misstep in contract documents.[2]

If the owner makes an improvident choice in environmental consultants, the owner may end up paying for that poor choice over and over: in price, in schedule, and in lost revenue when the project does not get completed on time.

The qualified and experienced environmental consultant will see to it not only that the contract documents protect the owner, but also that the owner has attended to the basics and that the contract documents are clear to the bidders. This approach should provide (1) a way of obtaining the best bids for the owner,

---

[2] Developing a carefully tailored agreement for each aspect of the renovation project is essential to controlling the differing site conditions claim. This is especially true with the design and implementation of the asbestos abatement project. Included are the owner-asbestos abatement contractor agreement, the owner-environmental consultant agreement, and the contractor-subcontractor agreement. Osborn & Schreyer, *Dealing with Environmental Risks in Construction Contracts,* Construction Specifier, July 1988, at 8; Osborn & Schreyer, *Asbestos Abatement Contractor Agreements,* Nat'l Insulation & Abatement Contractors Outlook Magazine, July 1989, at 29; Osborn, *Voice: Developing a Standard Form of Asbestos Agreement:* Nat'l Insulation & Abatement Contractors Outlook Magazine, Nov. 1989, at 84. The National Insulation and Abatement Contractors Association offers the NIAC Owner-Contractor Asbestos Abatement Agreement Guide (2d ed. 1990).

as the contractors will not be required to build in a contingency for potential misunderstandings in scope or intent, and (2) protection against being required to pay differing site conditions claims later.

The following clauses should be included in the business terms of the contract documents and carefully tailored to the project:

Scope of work. There is no doubt that the scope of work clause is the most important clause. It is clearly within the capability of the environmental consultant and the owner to draft this clause properly if they have done an adequate investigation of the site before writing it.

The owner can make foolish choices and use the scope of work clause as an attempt to prevent the contractor from getting paid for changed conditions if they are encountered. An example of such a clause is:

The Owner's Consultant has prepared a survey report which has been delivered to the Contractor. The Contractor acknowledges that the survey report has been received and reviewed. The Contractor shall remove *all Asbestos Containing Material in the building.* The Consultant's survey report is dated August 17, 1991 and identifies the Asbestos Containing Material anticipated to be encountered.

This clause causes trouble because it intends to obligate the contractor to take responsibility for asbestos removal beyond that which may have been contemplated by the consultant or the owner. It would appear to take the pressure off the consultant or owner to have found the areas where the asbestos existed.

It is likely that the court would construe this clause against the owner, especially if the asbestos is found in a wholly different area of the building than that shown on the abatement plan. However, the contractor should not feel comfortable relying on the equity of a court or the owner when seeking to obtain additional deserved compensation.

The good scope of work clause, together with well-defined plans and specifications, does two things:

1. It locates, describes, and quantifies all asbestos-containing material (ACM) in the building or facility.
2. It identifies the response action required to be taken for each item of ACM so designated.

Compliance with all laws. Defined correctly, a change in laws, regulations, or their enforcement, after a lump-sum asbestos abatement project has begun, will be deemed a differing site condition for which the contractor will receive additional compensation. Defined improperly, any such change will be a cost the contractor must bear.

A standard clause that is not helpful states:

The Contractor shall give notices and comply with all applicable laws, ordinances, rules, regulations and lawful orders of public authorities.

In order for the contractor to be protected by the clause, the following language must be added: "in effect and being enforced at the time of the signing of this agreement."

Sequencing of the work, suspension, and damages for delay. Lump-sum contracts for asbestos abatement may also include the following troublesome clauses:

The Contractor shall perform the Work in such order and sequence as directed by the Owner or the Owner's Consultant.

* * *

The Owner may, without cause, order the Contractor, in writing, to suspend, delay or interrupt the Work in whole or in part for such period of time as the Owner may determine.

* * *

The Contractor agrees to make no claim against the Owner for damages for delay or interference in the performance of this Agreement occasioned by any act or omission to act of the Owner or any of its representatives or for any other cause whatsoever, and agrees that any such claim shall be fully compensated for by an extension of time to complete the Work as provided herein.

On the asbestos abatement project, there can be no doubt that signing an agreement containing these clauses gives the owner carte blanche to disrupt, fail to give information, and fail to follow up on information and decision making. If faced with a decision as to whether to sign or lose the opportunity to do the job, it may be best to forego the job. If a contractor does sign a lump-sum contract with these clauses, it may lose a substantial sum when a differing site condition arises.

Contractor obligation to visit site. Many contracts contain a clause stating that the contractor has visited the site, has examined it before placing his bid, and is fully familiar with it. This type of clause may rule out a differing site conditions claim, although the courts will surely construe it against the owner if given the opportunity to do so. If the contract is negotiated, this clause should be clarified.

What is not clear is whether the contractor's inspection of the site includes any destructive testing or taking of samples. The clause should state the nature of the actual examination undertaken or required to be undertaken.

Differing site conditions clause. The following site investigation clause indicates that the full risk of a changed condition is on the contractor:

The Contractor represents that it has had the opportunity to examine and has examined carefully the plans and specifications and the contract and has acquainted itself with the conditions of the Work site; and that it has made all investigations essential to a full understanding of the difficulties which may be encountered in the performance of the terms of this agreement. The contractor will, regardless of any such conditions which may exist at the site of the project or its surroundings, assume complete responsibility for completion of the project under any such conditions which may exist at the site of the project or its surroundings and all risks in connection therewith. In addition thereto, the contractor represents that it is qualified fully and able to complete the project in accordance with the terms of the contract and within the specified time.

On the other hand, paragraph 4.3.6 from the AIA General Conditions of the Contract for Construction, AIA Document A201, 1987 edition, specifically allows for recovery for changed conditions claims.

When no asbestos is anticipated. Article 10 of AIA Document A201's General Conditions of the Contract for Construction, 1987 edition, includes a series of provisions that attempt to place the responsibility for detecting and dealing with asbestos on the owner.

The Engineers' Joint Contract Documents Committee (EJCDC) Standard General Conditions of the Construction Contract 1910-8 (1990) goes further than the AIA in affording protection to the contractor who does not anticipate finding asbestos. The applicable provisions read:

4.5.1 OWNER shall be responsible for any Asbestos, PCBs, Petroleum, Hazardous Waste or Radioactive Material uncovered or revealed at the site which was not shown or indicated in the Drawings or Specifications or identified in the Contract Documents to be within the scope of the Work and which may present a substantial danger to persons or property exposed thereto . . .

4.5.2. CONTRACTOR shall immediately: (i) stop all Work in connection with such hazardous condition and in any area affected thereby . . . CONTRACTOR shall not be required to resume Work in connection with such hazardous condition or in any such affected area until after OWNER has obtained any required permits related thereto and delivered to CONTRACTOR special written notice . . . .

4.5.3. If after receipt of such special written notice CONTRACTOR does not agree to resume such Work based on a reasonable belief it is unsafe, or does not agree to resume such Work stoppage under such special conditions, then Owner may order such portion of the Work that is in connection with such hazardous condition or in such affected area to be deleted from the Work.

§ 13.5 Avoiding Differing Site Conditions Claims

The primary way that an owner can avoid differing site conditions claims is not by drafting a better contract or hiring a better lawyer, but by doing its job better. The effect of a changed condition on the progress and cost of an abatement project may depend upon the owner's asbestos management philosophy. One of the most compelling arguments in favor of *total* asbestos removal is that it eliminates potential lawsuits and shields against future exposure. Although medical risk from low-level exposure may be extremely small, it is not nonexistent, as there is no minimum threshold for mesothelioma, an asbestos-associated cancer of the mesothelium or chest cavity lining.[3]

Although a worker who mined, manufactured, or installed asbestos is open to significantly greater risk, office workers' lawsuit are more than just a theoretical possibility. In *Layne v. GAF Corp.,*[4] an office worker who was employed from 1973 to 1985 as a word processor in a federal office building containing asbestos recovered from an asbestos manufacturer after she developed mesothelioma.[5]

§ 13.6 Analysis of Damages

Differing site conditions claim analysis and litigation, because it most often does not derive from a clear record, relies on careful review of contemporaneous records and the testing of witnesses' memories. Both liability and damages issues are decided in this fashion.[6]

Before the work begins, it may be possible to determine an acceptable formula by which the contractor will be paid if it is to remove more asbestos than is anticipated by the original contract or if the project is held up and the contractor, while doing the same work, will have to be at the job for longer. These suggestions apply to any construction project but become more essential in the asbestos abatement context because of the speed at which the contractor is working and the significant amount of labor expended each day.

The acceptable unit prices should be placed right in the contract so the parties can avoid the wrenching process of computing costs if doing so becomes necessary.

Types of damages the parties should be concerned with include:

1. Labor (extended, additional, idle, or escalated)
2. Material (additional or escalated)
3. Supervision in the field (field overhead) (extended or escalated)

---

[3] "Asbestos is a known human carcinogen for which no level of exposure is known to be without risk. Single exposures may even present a health risk to some individuals." Environmental Protection Agency (EPA) and National Institute of Safety and Health (NIOSH), A Guide to Respiratory Protection for the Asbestos Abatement Industry 2 (1986). Other asbestos-related diseases include asbestosis, an irreversible lung disease, and lung cancer.

[4] 1989 WL 155959 (Ohio Comm. Pl. 1989).

[5] *See A Legal Guide for Dealing with Asbestos in Buildings, in* Business Laws, Inc., ch. 41 (1989).

[6] Levin, *Costs Recoverable for Differing Subsurface Conditions, in* Hazardous Waste Disposal and Underground Construction Law 99 (R. Cushman & B. Ficken eds., John Wiley & Sons 1987).

4. Supervision from the home office (home office overhead) (extended, additional, idle, or escalated)
5. Equipment (extended, idle, or escalated).

If there is no predetermined unit price, it will be necessary to review records that show the increased costs of performance of the work and impact costs for the delay and disruption of other work activities resulting from the condition.

The records that must be reviewed are those showing the extent and rate of production for work activities affected by the condition and documenting crew and equipment usage in performing the work.

The factual confusion that can result from an asbestos differing site conditions claim is illustrated in *Universal Contracting & Brick Pointing Co. v. United States.*[7] In the *Universal* case, the court found that material issues of fact precluded summary judgment on the contractor's claim for extra compensation on (1) the claim that the owner furnished defective specifications; (2) whether presence of asbestos and glue constituted differing site conditions; and (3) whether the owner had superior knowledge of presence of asbestos and glue in the paint system.

Count III of the contractor's complaint alleged that, in performing the contract, the contractor encountered asbestos on the exterior walls of the buildings, which constituted a differing site condition under the terms of the contract, and that this condition was not discoverable through a reasonable site inspection.

The contract contained a clause covering differing site conditions, which stated in relevant part:

> (a) The Contractor shall promptly, and before such conditions are disturbed, notify the Contracting Officer in writing of:
>
> (1) subsurface or latent physical conditions at the site differing materially from those indicated in this contract, or
>
> (2) unknown physical conditions at the site, of an unusual nature, differing materially from those ordinarily encountered and generally recognized as inhering in work of the character provided for in this contract.[8]

The contractor claimed that the presence of asbestos in the buildings satisfied the elements of either differing site condition category. With respect to category 1, the contractor argued that the contract documents indicated that the contractor would be required to remove paint, not asbestos, in performing the contract, and that the contract gave no indication that such a latent physical condition as asbestos would be encountered beneath the layers of paint. The contractor contended that the site conditions, as indicated at several places in

---

[7] 19 Cl. Ct. 785 (1990).

[8] *Id.* at 795.

the contract documents, gave rise to a permissible inference that only paint would be encountered during contract performance.

With regard to category 2, the contractor argued that asbestos is not a material ordinarily encountered in paint removal contracts, and that no other contractor at the site encountered this type of material.

In *Owners Realty Management Construction Corp. v. Board of Education,*[9] at the request of the architect and assistant supervisors, the contractor, under protest, removed asbestos in the owner's schools from areas the contractor claimed were outside the scope of the contract.

The contractor's summary judgment motion was denied because there were triable issues of fact as to whether the work ordered by the owner's representatives constituted such an extraordinary demand that there "'could be no reasonable doubt that [the demand] exceeded the obligations of the contract and that a refusal to comply with it would not work a breach of contract.'"[10]

The court concluded that the asbestos removal performed by the contractor at the owner's representatives' directions was "so preposterous, so manifestly beyond the limits of the agreement, as to suggest a threat to the public interest."[11]

In a 1990 case, a contractor was performing under a temperature and humidity control contract for rehabilitation work in the main branch of the New York Public Library at Fifth Avenue and 42nd Street in Manhattan. Eight months after work began, the owner (the city of New York) for the first time said that the insulation surrounding some existing piping and ductwork was composed of hazardous asbestos, which would have to be removed.[12]

After extensive negotiations and correspondence, the contractor refused to undertake the work in compliance with a change order issued by the city because the work to be completed pursuant to the change order exceeded the 5-percent limit,[13] funding had not been approved, and the city would not hold the contractor harmless from any asbestos-related claims.

The city defaulted the contractor. The lower court held that the contractor was in default and that the surety had waived the claim that the asbestos removal change order constituted a material change of the contract by negotiating the price of the work with a subcontractor and submitting an estimate of the cost of the work on the change order.

---

[9] 160 A.D.2d 921, 554 N.Y.S.2d 648 (1990).

[10] *Id.* at 921, 554 N.Y.S.2d at 648 (quoting Kalisch-Jarcho, Inc. v. City of N.Y., 72 N.Y.2d 727, 736, 533 N.E.2d 258, 263, 536 N.Y.S.2d 419, 424 (1988) (quoting Borough Constr. Co. v. City of N.Y., 200 N.Y. 149, 158 (1910))).

[11] Kalisch-Jarcho, Inc. v. City of N.Y., 72 N.Y.2d 727, 736, 533 N.E.2d 258, 263, 536 N.Y.S.2d 419, 424 (1988)).

[12] Aetna Casualty & Sur. Co. v. City of N.Y., 160 A.D.2d 561, 554 N.Y.S.2d 210 (1990).

[13] The New York City Administrative Code provides that whenever a contract is increased by the municipal agency by more than 5% of the original contract amount, Board of Estimate approval is required.

The appellate court decided that the contractor and surety could relitigate the default and were not barred by res judicata for the default proceeding.

Tuscaloosa City Board of Educ. v. American/Owens, Inc.[14] involved litigation over the method of dispute resolution. Under the contract, American/Owens agreed to remove asbestos-containing material from buildings owned by the board of education for a lump sum, subject to additions and deductions provided for in the contract.

The contract provided that "[a]ny dispute, claim, or question concerning the interpretation of meaning of the Contract Documents, or concerning a breach of the Contract, shall be submitted to the Director and his decision shall be final and binding on the parties to the Contract."[15]

After beginning the work, American/Owens reported to the board that the quantity of asbestos to be removed from the buildings exceeded the amount it expected to find based on drawings and specifications made by the architect employed by the board, and that the work would be done as an extra item not within the contract.

When the board refused to pay for extra work, a hearing before the director of the State Building Commission was convened, at which the board argued the director lacked jurisdiction to resolve the controversy. Thereafter, the director entered an order requiring payment of the additional amount demanded by American/Owens.

The board filed a declaratory judgment action, claiming that the dispute was not subject to resolution by the director on the basis that only disagreements over "contract provisions" were within the settlement authority of the director.

The court found the contract provisions to contain inconsistent or unclear provisions that the arbitrator interpreted in resolving the dispute. These contract provisions are worthy of note here as a lesson for owners who wish to avoid a costly dispute resolution process by focusing, up front, on contract drafting. The contract provisions were:

Scope of Work

The contractor will be responsible for the removal of all friable asbestos in student occupied areas, including, but not limited to, that covered in the base bid. In the event there is a question as to whether material is friable, or whether the area is student occupied, the architect/engineer will make the final decision.

* * *

General Conditions of the Contract

Changes in the Work

The Owner may at any time make changes in the Work by changes in the Drawings and Specifications of the Contract and within the general scope

---

[14] 486 So. 2d 405 (Ala. 1986).

[15] *Id.* at 406.

thereof. Changes will be in the form of a Contact Change Order based upon a written request of the Owner and a written proposal of the Contractor. All Change Orders will require consent of Surety by endorsement of the Change Order form. In making any change, the charge or credit for altering, adding to or deducting from the Work shall be determined by one of the following methods selected by the Owner:

* * *

Claims for Extra Cost or Extra Work

If the Contractor claims that any instructions by drawings or otherwise are not in accordance with the Contract Documents, and involve extra cost under the Contract, he shall give the Architect and the Director written notice thereof within ten (10) days after receipt of such instructions, and in any event before proceeding to execute the work and the procedure shall then be as above under [the Changes in Work section]. Otherwise no claim will be considered.[16]

The case of Castle Construction Co./Tuskegee Lumber Co. v. Owens & Woods Partnership[17] demonstrates the treatment of the encountering of asbestos after a renovation has begun.

Castle Construction Company/Tuskegee Lumber Company (Castle) entered a contract to renovate and modernize the Central City Housing Project in Birmingham, Alabama, owned by the Housing Authority of the Birmingham District (HABD).

Castle suspended work when unanticipated asbestos was encountered in each of the buildings being renovated. Castle advised the architect that delay would result. The architect responded by changing the project specifications to provide for the asbestos removal, and the owner, the architect, and the contractor agreed to two change orders that afforded the contractor with the direct costs of the asbestos removal.

After the completion of the project, the contractor presented a claim for delay costs suffered in addition to the direct costs for the asbestos removal. When HABD denied the claim, the contractor sued for breach of contract.

The contractor sought increased costs it suffered as a result of the asbestos-related delay, plus similar costs its subcontractors incurred.

The court stated:

> The discovery of asbestos was clearly a 'latent defect' as defined in the contract, and the change orders entered into between HABD and Castle compensated Castle and its subcontractors for the additional labor and materials required for its removal. Accordingly, we also hold that Castle is not entitled to delay or disruption damages from HABD or Owens and Woods on behalf of its subcontractors.[18]

---

16 *Id.*

17 1991 WL 184478 (Ala. Aug. 30, 1991).

18 *Id.*

The contract provisions in this case were as follows:

Changes in the Work

8(b) Any change in the work shall be ordered and the adjustment of the contract price or time shall be determined by one of the following methods: Method 1—Adjustment before Performance . . . Method 2—Adjustment after Performance . . . The Contractor shall keep an accurate current account of such work and present it in such form, and substantiated by such supporting papers as the Architect may require. Upon completion and determination of the cost, a Change Order shall be issued establishing the increase or decrease in the contract price or contract time. . . .

8(e) The cost shall include all direct and necessary production costs of the work itself . . .

9(b) Should the Contractor encounter subsurface or latent conditions at the site materially differing from those provided for in this Contract, or unknown physical conditions differing from those inherent in work of the character provided in this Contract, he shall promptly, and before such conditions are disturbed, notify [the Owner] in writing . . .

9(e) If, on the basis of available evidence, [the Owner] determines that an adjustment of the contract price or time is justifiable, the procedure shall then be as provided herein for 'changes in the work.'[19]

§ 13.7 Strategy for Contract Administration and Claim Management

Owner's Strategy

The owner's claims prevention and claims management strategy must start on the first day of ownership of the building. The owner must develop a comprehensive plan to manage asbestos. The owner's asbestos management program, if set up properly, becomes the owner's contract administration and claims management program. Whenever a move is made in the interior of the building containing asbestos which is being maintained in place, the owner's overall asbestos management plan must be consulted.

In developing an asbestos management policy, owners should follow these 10 steps:

1. Take a detailed inventory of all asbestos in the building, its condition, and what is being done about it. Keep this inventory updated, and make sure it is reviewed before any renovation is undertaken. If sued, the building owner is not "off the hook" merely because the owner did not know of an

---

[19] *Id.*

asbestos problem; the building owner is held to the standard of "what a reasonably prudent owner in the locality should have known under the circumstances."[20] In current terms, there is much the owner "should have known" about asbestos.

2. Once an inventory shows the presence of asbestos, prepare a written operations and maintenance (O&M) plan, and make sure that all relevant personnel are familiar with it. Make sure that the O&M plan is kept up-to-date.
3. Hire a facilities management staff carefully and make sure they are fully aware of all aspects of asbestos maintenance and removal. Make sure they attend formal training on an ongoing basis.
4. Set up a consistent and well-organized recordkeeping system relating to asbestos management, and monitor it to assure that it is being followed.
5. Formulate a written plan for addressing emergencies relating to in-place asbestos (such as a burst heating pipe or a water leak), and make sure that facilities management personnel are fully trained in these procedures. It is important that emergencies be addressed promptly and thoroughly and that proper contemporaneous documentation be kept. The 24-hour phone numbers of the asbestos management team should be made available to facilities management personnel and executive level personnel and the owner's engineer, lawyer, and publicist. It is important that a chain of command be identified and that a coordinator be appointed for this important function.
6. Develop a hands-on program; know firsthand the quality of the asbestos management program. Do not rely solely on the advice of others. If you are an owner or an executive of the owner, know all of the facts and ask questions.
7. Do not rely on insurance. Expense, policy limits, sunset clauses, exclusions, the claims-made nature of coverage, the long-delayed manifestation of asbestos-related liability, and questionable long-term stability of insurance companies preclude such reliance. If an abatement is being conducted, make sure that the contractor obtains asbestos coverage on an occurrence basis from a stable and rated insurer.
8. Do not rely on contract clauses that purport to provide protection. Indemnity clauses or other assurances provided by prior owners, abatement contractors, transporters, or others are only as good as their future solvency. However, even though it is not a complete answer, make sure that proper protection is obtained when a purchase, sale, or lease is entered or when an abatement contractor is hired.
9. Worry about where removed asbestos is being disposed. If the asbestos is removed and improperly disposed, the owner may be sued to clean up the

[20] Restatement (Second) of Torts § 299A.

dump site years later. Although the asbestos in the building is not subject to the Superfund laws,[21] once it is incorporated into the dump site, it is subject to the Superfund and other strict liability laws. The owner could then be responsible for a multimillion-dollar cleanup even though the owner did not dispose of the asbestos.

10. Be candid about the presence of asbestos with tenants, potential purchasers, regulatory authorities, and the public. Formulate an "enlightened" public relations plan and follow it. The worst legal exposure is often due not to the presence of asbestos, but to the lengths that are taken to conceal its presence. Be careful: perceptions, especially in the case of an emergency, may govern. A panicky response leads to improper handling. Instruct the team to be calm and to avoid guessing or speculating about the facts.[22]

Public opinion and media reports have led some building owners to believe that asbestos is no longer dangerous. The skepticism about devoting resources to proper management of in-place asbestos was created by reports and forums, beginning in December 1988 with a symposium held at Harvard University called the "International Symposium on the Health Aspects of Exposure to Asbestos in Buildings." The Harvard symposium, sponsored by many real estate industry groups, produced a report released in August 1989 branding the public's concern with asbestos as "fiber phobia."[23] The report concluded that risks posed by in-place asbestos are very low and suggested that money spent for asbestos abatement would be better spent on abating greater risks like cigarette smoking. Subsequent articles were published in the *New England Journal of Medicine*[24] and *Science Magazine.*[25]

---

[21] 42 U.S.C. §§ 9607–9675 (1988). The Comprehensive Environmental Response, Compensation, and Liability Act (CERCLA), also known as Superfund, was enacted in 1980 and revised by the Superfund Amendments and Reauthorization Act of 1986 (SARA), Pub. L. No. 99-499, 100 Stat. 1613 (1986). The principal purpose is to provide a mechanism for public and private cleanup of hazardous waste sites. Note that asbestos as it is in place inside buildings is not hazardous waste, but once removed and placed in a waste disposal site, asbestos is covered by the Superfund laws.

[22] These steps were first presented to an owner-oriented audience at the New York University Real Estate Institute's seminar, *Asbestos in New York City Buildings* (1990). The guidelines were later published in Osborn, *New York City's Proposed In-Place Asbestos Inspection Law: What It Means for Building Owners, in* The 1991 Asbestos Sourcebook, and later dubbed "The Ten Asbestos Commandments."

[23] Harvard University, *International Symposium on Health Aspects of Exposure to Asbestos in Buildings,* Aug. 1989.

[24] New England Journal of Medicine, June 29, 1989.

[25] Science Magazine, Jan. 19, 1990. Although the *New York Times Magazine* in its November 25, 1990, issue (page 39) offers a well-reasoned discussion of the *Science Magazine* and *New England Journal of Medicine* articles, the title of the article could be somewhat misleading: "The Asbestos Mess: Now Some Scientists Say Removing the Fiber Can Be Worse Than Leaving It."

The *Science Magazine* article, in its section entitled "Public Policy," stated:

> [T]he available data and comparative risk assessments indicate that chrysotile asbestos, the type of fiber found predominantly in U.S. schools and buildings, is not a health risk in the non-occupational environment. Clearly the asbestos panic in the U.S. must be curtailed, especially because unwarranted and poorly controlled abatement results in unnecessary risks to removal workers who may develop asbestos related cancers in later decades.[26]

Based on this information, these headlines appeared:

"Study Opposes Removal of Asbestos"

"Health Risks of Asbestos Downplayed"

"Risk Is Seen in Needless Removal of Asbestos"

"Health Experts Say Billions May Be Wasted by Removing Banned Insulation Material"

The building owner should not change the long-range asbestos policy set out in the "Ten Asbestos Commandments." Unless asbestos has become airborne through use or one is about to disturb it, there is no requirement that an owner remove it. In fact, the Environmental Protection Agency (EPA) has issued guidelines encouraging owners to set strategies to manage asbestos in place by using operations and maintenance (O&M) plans.[27]

Probably imprudently, "in-place" legislation, which would have required asbestos inspections in buildings, was shelved, locally in cities such as New York, and in Congress. These laws did not require removal of asbestos. Their purpose was to "create a level playing field" for building owners and to endorse a "responsible asbestos policy" to which owners should adhere. They would simply have required that the buildings be inspected to see if there was an asbestos problem, and if there was a problem, that a solution be formulated and implemented.

Note that when a 1988 study was conducted by the city of New York of 900 buildings there, 84 percent had asbestos in poor or fair condition. Whether there is a statutory requirement or not, the building owner is responsible for properly managing the asbestos in place or for removing it. The study further showed that two-thirds of all buildings in New York City contained asbestos containing

---

[26] *Id.*

[27] Osborn, *Perspectives on O&M,* Nat'l Insulation & Abatement Contractors Outlook Magazine, July 1991, at 34, and EPA, The Asbestos Green Book: Managing Asbestos in Place (1990). As the EPA Green Book states:

> There are steps which a building owner can take to prevent asbestos fiber releases or resuspensions of already-released fibers, or control fiber releases quickly and safely if they occur. O&M [operations and maintenance] programs are designed to achieve both these goals.

material, and 80 percent of all office buildings above seven stories contained asbestos.

Further reports on asbestos's health effects were published in the fall of 1991.[28] Such reports and studies should not lull the responsible building owner into ignoring proper management of asbestos.[29]

It is important to hire a well-qualified environmental consultant. Without an environmental consultant to guide the asbestos management program, differing site conditions claims may still abound. This is especially true when the owner is following an in-place management program, which means there will be an array of small asbestos abatement jobs going on throughout the building as tenants vacate or as problems arise.

Beyond hiring the appropriate consultant, there must be a commitment to utilizing the consultant properly. This means that the consultant must be provided with full information about the building, including any changes in use, in leasing arrangements, or in management of the building. Once moves are made without keeping the environmental consultant advised (such as repairing sprinkler heads or fixing leaky pipes), the consultant's usefulness in helping to combat changed conditions claims will diminish.

Another aspect of appropriate utilization is in allowing the consultant to call for destructive testing or further investigation when deemed appropriate. Pennies spent at the outset to pinpoint a scope of work have saved tens of thousands of dollars that would have been spent to address and work around suspensions in work and delays to completion caused by failure to know site conditions.[30]

---

[28] In September 1991, both the American Medical Association and the Health Effects Institute issued studies recommending against full-scale removal of asbestos in buildings.

[29] EPA & NIOSH, A Guide to Respiratory Protection for the Asbestos Abatement Industry 2 (1986).

[30] The importance of the environmental consultant's role is discussed in Osborn, *The Owner-Environmental Consultant Agreement, in* Construction Renovation Formbook 163 (R. Cushman ed., John Wiley & Sons 1991). Further, in March 1991, a report entitled "Report of the Judicial Conference, Ad Hoc Committee on Asbestos Litigation," was issued to the Chief Justice of the United States Supreme Court. The report concluded that the resolution of asbestos personal injury cases is "an impending disaster" deriving from an estimated 200,000 asbestos disease deaths by the year 2000 or 265,000 by the year 2015. In a court system with increased filings, larger backlogs, higher cost, and more bankruptcies, prospects are poor, said the report, that judgments, when obtained, can be satisfied, as financial resources of the manufacturers and their insurance policies become fully depleted. *See* Buckley & Co. v. City of N.Y., 505 N.Y.S.2d 140 (App. Div. 1986). In this case, the New York State Appellate Division upheld the decision of Justice Martin Evans, which held that a differing site condition was contemplated by the parties because there was a clause in the contract covering differing site conditions. When read together with a clause stating that no damages were to be allowed for delay in the project, no "impact" or "delay" cost was recoverable.

Contractor's Strategy

The contractor's strategy, of course, is not set from as early a point as that of the owner. Although the owner has been able to plan and prepare its approach from the time the building was owned, the contractor most often starts planning at the time it bids on the contract. Most often, the contractor is not able to set the contract terms and may not, in the case of a sealed bid, be able to negotiate them. In addition, there may be a number of clauses that will prevent the implementation of a successful strategy for obtaining prompt payment on a differing site conditions clause.

The contractor's leverage is never greater than when the owner needs the contractor. The owner needs the contractor *before* it has fully performed and *before* it becomes easy to kick the contractor out and bring in another one.

It is clear that the clause requiring the contractor to keep working while a dispute on changed conditions is resolved is harmful to the contractor's leverage rights. If the owner must face a situation in which job progress is affected, the owner will "light a fire" under the contractor because the contractor will be in default and the owner will be able to bring in a replacement if the contractor fails to continue.

Leverage is a tricky mechanism, and it is not a good part of contractor strategy for the contractor to walk off the job or to suspend work in the face of a clause that calls for its continued work during the resolution of a dispute.[31] If enough money is involved, there must be an interim solution of some sort if the contractor is to keep meeting payroll.[32] Perhaps there is a way for the contractor to obtain payment to the extent the owner feels appropriate and to reserve the right to sue for the rest later. It is to be expected that if an interim payment is to be made by the owner, that the owner will expect a full release for that item; that is the owner's use of his leverage—after all, the contractor needs the payment now and maybe to get it he will take less.

§ 13.8 Alternative Dispute Resolution

Many differing site conditions claims involving management of asbestos have been settled without resort to litigation, perhaps because of both owners' and contractors' relative inexperience in asbestos abatement. The reason could, of course, simply be that owners would prefer not to see themselves locked in disputes over their environmental problems in a public forum.

---

[31] *See, e.g.*, AIA A201-1987, Standard Form of Agreement Between Owner and Contractor ¶ 4.3.4.

[32] *See* American Society of Civil Engineers, Avoiding and Resolving Disputes in Underground Construction: Successful Practices and Guidelines (1989). The publication is available from the ASCE, 345 East 47th Street, New York, N.Y. 10017-2398.

The 1980s saw the passage of a massive volume of new asbestos-related legislation and regulations and the development of common law doctrines that broaden existing areas of legal responsibility relating to asbestos.

As a result, for the 1990s, the federal and state courts are in for an onslaught of new cases, including asbestos changed conditions cases. Many of them, in volume of documents, complexity of legal and factual issues, number of witnesses, and number of parties involved, will resemble antitrust, construction contract, and other large and protracted commercial litigation that already confounds the courts.

Clearly, the alternatives to the time-consuming and costly conventional means of dispute resolution may be welcome in this heavily regulated industry. Alternative dispute resolution (ADR) has been used with great success in commercial and contractual disputes.[33] It is clear that with the heavy docket of the courts, commercial litigation will take many years to reach trial, although 90 to 95 percent of all lawsuits are settled before trial. The goal, then, must be to replicate, at an earlier stage, the same types of dynamics that are present on the courthouse steps.[34]

Proponents argue that ADR has many advantages:

1. It accelerates the dispute resolution process, which allows the parties to focus on their interests rather than on the procedural strategy of the case, and provides "a day in court," allowing for advocacy and emotional outlets.
2. It assures the involvement of executives early in the process.
3. It avoids the need to educate a judge or jury.
4. It can better protect confidentiality.
5. It avoids acrimony that often accompanies litigation and allows for a business relationship between the parties to continue.

In a situation in which the parties enter a contract with each other, such as one between an owner and asbestos abatement contractor, it is appropriate to specify, in the contract itself, the procedure under which any disputes between the contracting parties are to be resolved. Even though some parties feel there could not

---

[33] Careful drafting of the dispute resolution clause is essential and must answer questions such as: Where is the dispute to be venued? Which state's laws should be applied? Should interim disputes be initially determined by the architect or the environmental consultant? *See* Forster, *Broad Form Arbitration Clauses Create Confusion,* Consulting Specifying Eng'r, Oct. 1989, at 33. Should terms of another contract be deemed incorporated by reference? *See* Bigge Crane & Rigging Co. v. Docutel, 371 F. Supp. 240 (E.D.N.Y. 1973); J. Sweet, Sweet on Construction Industry Contracts, ch. 8 (John Wiley & Sons 1987). Do not be surprised if each clause has a whole body of case law construing it.

[34] Osborn, *Alternate Dispute Resolution and Asbestos,* Nat'l Insulation & Abatement Contractors Outlook Magazine, Apr. 1990, at 48; Grad, *Alternate Dispute Resolution in Environmental Law,* 14 Colum. J. Envtl. L. 157 (1989).

possibly be a problem and are unwilling to give attention to the inclusion of an ADR clause in their agreement, others, who have seen the economy created through the use of ADR, are now including ADR in the contract documents.

Certainly it is never too late to use ADR at any point, especially in cases that have been at an impasse after a number of years of traditional litigation. Impasse can arise from personality conflicts between counsel, poor communications between parties, inflexible negotiating postures, or difficult policy or internal political reasons.

The parties can rest assured that the court will encourage their use of ADR. In fact, the courts have increasingly suggested resort to ADR, especially in connection with technically complex issues that the court is ill suited to address. Changed conditions issues on asbestos projects surely fall into this category.

In contractual relationships, the method of dispute resolution can be a point of negotiation along with the other terms of the agreement. Approaches to certain recurring problems can be fashioned (such as scope of work disputes, evaluation of undisclosed conditions claims, and interpretation and application of government regulations).

There appears to be no single universal process that will work for every dispute. A number have been used, including:

1. **Mediation:** Facilitates settlement by use of a neutral party who has no power to make binding decisions, but who may schedule and structure negotiations, act as a catalyst between the parties, focus the discussions, aid communications, and serve as an assessor and not a judge of the positions of the parties.
2. **Arbitration:** Employs a neutral party to hear all or a portion of the factual, legal, or policy issues in a case and make binding decisions. Arbitration is generally less formal than court proceedings, as discovery is, in theory, severely restricted and the rules of evidence are not strictly followed.
3. **Fact finding:** Uses a neutral third party with industry expertise to narrow factual issues. The process can be either binding or nonbinding, and the parties, if they so choose, can treat the decision on the issues of fact as admissible in subsequent proceedings.
4. **Minitrial:** Permits parties to present their cases (or portions of their cases) to principals who have authority to settle the dispute. Following the presentation, the principals continue negotiations with the aid of the third party who has acted as the minitrial judge.[35]

---

[35] Following is a streamlined formula for implementing the minitrial agreement:

1. The parties designate a neutral advisor, usually a person of eminence or one highly knowledgeable in the subject matter of the dispute.
2. Principals of each party, with full settlement authority, attend the proceedings.

5. **Settlement judge:** Involves the participation of judges of the court hearing the case who are not involved in any aspect of the litigation. The settlement judge hears the positions of the parties and offers suggestions as to the reasonable compromises in light of the judge's evaluation of likely outcomes.
6. **Summary jury trial:** Involves an abbreviated trial before a jury assembled by the parties to get a sense of the amount likely to be awarded after a full trial. The decisions of the summary jury are advisory only.

No matter what type of lawsuit is at hand on an asbestos abatement project, it is likely to include the analysis of voluminous technical evidence. Therefore, exploration of ADR is wise. The alternative is waiting years for a resolution, with perhaps the same result, after expending tens and perhaps hundreds of thousands in attorneys' fees and court costs and an equally large amount of time. Under these circumstances, ADR becomes an attractive option.

---

3. Limited discovery is afforded both parties.
4. Evidence is submitted to the neutral advisor in advance of the hearings.
5. Each party is given a fixed, short period of time to make its presentation. There are no rules governing what evidence can be presented. No objections to evidence are permitted. No cross-examination is allowed, although the neutral advisor may ask questions.
6. The neutral advisor makes a written recommendation that the parties use as a basis for settlement negotiations.
7. If the case cannot be settled, the neutral advisor's recommendation is not disclosed in subsequent litigation or arbitration.
8. Sanctions are provided against the nonprevailing party if the ultimate result of arbitration or litigation is close to the neutral advisor's recommendations.

This formula appeared in Osborn, *Alternate Dispute Resolution and Asbestos,* Nat'l Insulation & Abatement Contractors Outlook Magazine, Apr. 1990, at 48, adapted from Rubin & Banick, *Alternative Dispute Resolution Forms, in* 2 Construction Industry Forms 361 (R. Cushman & G. Blick eds., John Wiley & Sons 1988).

TABLE OF CASES

Case	*Book §*
Construction Pan-Caribe, S.A., ENGBCA No. PCC18, 73-2 B.C.A. (CCH) ¶ 10,238 (1973)	§ 9.7
Continental Drilling Co., ENGBCA No. 3455, 75-2 B.C.A. (CCH) ¶ 11,541 (1975)	§§ 2.4, 4.7
Cooper v. Jevne, 128 Cal. Rptr. 724 (Ct. App. 1976)	§§ 6.2, 6.5
Corinno Civetta Constr. Corp. v. City of N.Y. (Honeywell), 493 N.E.2d 905 (N.Y. 1986)	§ 1.5
County Asphalt v. State, 337 N.Y.S.2d 415 (App. Div. 1972)	§ 4.7
COVCO Haw. Corp. ASBCA No. 26901, 83-2 B.C.A. (CCH) ¶ 16,554 (1983)	§§ 8.10, 8.11
Crook v. Oklahoma Bd. of Pub. Affairs, 736 P.2d 140 (Okla. 1987)	§ 5.4
CSH Contractors, Inc., NASA BCA No. 476-2, 77-2 B.C.A. (CCH) ¶ 12,814 (1977)	§ 11.3
C.W. Bignold v. King County, 399 P.2d 611 (Wash. 1969)	§ 3.3
C.W. Blakeslee & Sons v. United States, 89 Ct. Cl. 226 (1939), *cert. denied,* 309 U.S. 659 (1940)	§ 2.1
D. Federico Co. v. Commonwealth, 415 N.E.2d 855 (Mass. App. Ct. 1981)	§§ 4.4, 4.6
D. Federico Co. v. New Bedford Redev. Auth., 723 F.2d 122 (1st Cir. 1983)	§§ 3.3, 3.4, 4.9, 8.3
Daniel Hamm Drayage Co. v. Waldinger Corp., 666 F.2d 1213 (8th Cir. 1981)	§ 6.2
Darwin Constr. Co., ASBCA No. 27596, 86-1 B.C.A. (CCH) ¶ 18,645 (1985)	§ 11.9
Davidson & Jones, Inc. v. County of New Hanover, 255 S.E.2d 580 (N.C. Ct. App. 1979)	§ 4.9
Davidson & Jones, Inc. v. North Carolina Dep't of Admin., 337 S.E.2d 463 (N.C. 1985)	§ 3.3
Dawco Constr., Inc., ASBCA No. 31990, 88-2 B.C.A. (CCH) ¶ 20,606 (1988), *aff'd,* 867 F.2d 615 (Fed. Cir. 1989)	§§ 8.11, 11.6
Dawco Constr., Inc. v. United States, 930 F.2d 872 (Fed. Cir. 1991)	§ 6.3
Del E. Webb Constr. v. Richardson Hosp. Auth., 823 F.2d 145 (5th Cir. 1987)	§ 1.6
Delphcon/J.A.M.E., J.V., PSBCA No. 1455, 88-3 B.C.A. (CCH) ¶ 21,107 (1988)	§ 11.3
Diamond Pac., NASA BCA No. 45-0391, 92-1 B.C.A. (CCH) ¶ 24,615 (1991)	§ 11.6
Dillon v. United States, 156 F. Supp. 719 (Ct. Cl. 1957)	§ 2.3
Dobson v. State of N.J., 526 A.2d 1150 (N.J. Super. Ct. 1987)	§ 1.5
Dravo Corp. v. Litton Sys., 379 F. Supp. 37 (D. Miss. 1974)	§ 5.1
Dravo Corp. v. Seattle, 484 P.2d 399 (Wash. 1971)	§ 11.3
Eastern Tunneling Corp v. Southgate Sanitation Dist., 487 F. Supp. 109 (D. Colo. 1979)	§§ 1.4, 4.4, 4.6, 4.7, 9.5

Case	*Book §*
Gerald Miller Constr. Co., IBCA No. 2292, 91-2 B.C.A. (CCH) ¶ 23,829 (1991)	§ 8.5
Glenn's Heating, ASBCA No. 25754, 83-1 B.C.A. (CCH) ¶ 16,358 (1983)	§ 8.3
Glynn v. City of Gloucester, 401 N.E.2d 886 (Mass. 1980)	§ 3.3
Gogo v. Los Angeles County Flood Control Dist., 114 P.2d 65 (Cal. Ct. App.), *motion denied,* 117 P.2d 392 (Cal. Ct. App. 1941)	§ 6.2
Golomore Assocs. v. New Jersey State Highway Auth., 413 A.2d 361 (N.J. Super. Ct. App. Div. 1980)	§§ 4.5, 6.1, 9.5
Granite Constr. Co., IBCA No. 1500-8-81, 82-1 B.C.A. (CCH) ¶ 15,631 (1982)	§ 11.3
Granite-Groves v. Washington Metro. Transit Auth., 845 F.2d 330 (D.C. Cir. 1988)	§ 6.5
Great Lakes Dredge & Dock Co., ENGBCA No. 5606, 91-1 B.C.A. (CCH) ¶ 23,613 (1990)	§§ 8.10, 8.12
Grow Constr. Co. v. State, 391 N.Y.S.2d 726 (App. Div. 1977)	§§ 6.3, 6.5
Gust K. Newberg, Inc. v. Illinois State Toll Highway Auth., 506 N.E.2d 658 (Ill. App. Ct. 1987)	§ 1.5
Hallman v. United States, 68 F. Supp. 204 (Ct. Cl. 1946)	§ 8.3
Hallmark Elec. Contractors, Inc., ASBCA No. 32595, 87-2 B.C.A. (CCH) ¶ 19,870 (1987)	§ 8.8
Halstead Oil Co. v. Northern Ins. Co., No. 7043/90 (N.Y. 1990)	§ 12.12
Hardaway Contractors, Inc., 3 MSBCA ¶ 227 (1989)	§ 9.3
Hardeman-Monier-Hutcherson v. United States, 458 F.2d 1364 (Ct. Cl. 1972)	§§ 2.3, 8.3, 9.7, 11.5
Helene Curtis Indus., Inc. v. United States, 160 Ct. Cl. 437, 312 F.2d 774 (1963)	§ 4.8
Henshaw v. Edward E. Clark Eng'rs-Scientists, Inc., 490 So. 2d 161 (Fla. Dist. Ct. App. 1986)	§ 6.5
Hercules Constr. Co., VABCA No. 2508, 88-2 B.C.A. (CCH) ¶ 20,527 (1988)	§ 8.10
Herman H. Neumann, IBCA No. 6821167, 70-1 B.C.A. (CCH) ¶ 8216 (1970)	§ 9.7
Hersey Gravel Co. v. State Highway Dep't, 9 N.W.2d 567 (Mich. 1943)	§ 6.3
Highland Constr. Co. v. Stevenson, 636 P.2d 1034 (Utah 1981)	§§ 5.4, 9.4
Highland Reforestation, Inc. v. United States, 230 Ct. Cl. 1020 (1982)	§ 8.11
Hollerbach v. United States, 233 U.S. 165, 34 S. Ct. 553 (1914)	§§ 4.3, 4.6, 4.9, 6.6
Holloway Constr. Co., ENGBCA No. 4805, 89-2 B.C.A. (CCH) ¶ 21,713 (1989)	§ 11.3
Horton Indus. Inc. v. Village of Moweaqua, 492 N.E.2d 220 (Ill. App. Ct. 1986)	§ 4.8
Hoyer Constr. Co., ASBCA No. 21616, 84-2 B.C.A. (CCH) ¶ 17,249 (1984)	§ 8.9

Case	*Book §*
Hurlen Constr. Co., ASBCA No. 31069, 86-1 B.C.A. (CCH) ¶ 18,690 (1985)	§§ 8.9, 8.10
Husman Bros., DOT CAB No. 71-15, 73-1 B.C.A. (CCH) ¶ 9889 (1973)	§ 11.5
Husman, Inc. v. Triton Coal Co., 809 P.2d 796 (Wyo.), *reh'g denied,* 191 Wyo. LEXIS (Wyo. 1991)	§§ 4.8, 4.9
Illinois Constructors Corp., ENGBCA No. 5678, 91-2 B.C.A. (CCH) ¶ 23,728 (1991)	§§ 8.8, 8.9
Imbus Roofing Co., GSBCA No. 10430, 91-2 B.C.A. (CCH) ¶ 23,820 (1991)	§§ 5.3, 8.8, 8.10, 8.12
Indianapolis, City of v. Twin Lakes Enter., Inc., 568 N.E.2d 1073 (Ind. Ct. App. 1991)	§ 11.5
Industrial Dredging & Eng'g Corp. v. Southern Ind. Gas & Elec. Co., 840 F.2d 523 (7th Cir. 1988)	§ 4.9
Inland Bridge, Inc. v. North Carolina State Highway Comm'n, 227 S.E.2d 648 (1976)	§ 3.4
Instalaciones Aire Frio S.A., ENGBCA No. PCC-44, 85-2 B.C.A. (CCH) ¶ 18,015 (1985)	§ 8.10
Intel Corp. v. Hartford Accident & Indem. Co., 952 F.2d 1551 (9th Cir. 1992)	§ 12.12
IT Corp. v. MOTCO Site Trust Fund, No. 91-3532 (S.D. Tex. Dec. 3, 1991)	§ 2.5
J. Lawson Jones Constr. Co., ENGBCA No. 4363, 86-1 B.C.A. (CCH) ¶ 18,719 (1986)	§ 11.3
J. Parr Constr. & Design, Inc. v. United States, 24 Cl. Ct. 228 (1991)	§ 11.3
J.A. LaPorte, Inc., ENGBCA No. 5254, 88-3 B.C.A. (CCH) ¶ 20,953 (1988)	§ 8.3
J.A. Terteling & Sons, Inc. v. United States, 390 F.2d 926 (Ct. Cl. 1968)	§ 9.4
J.A. Thompson & Son, Inc. v. State, 465 P.2d 148 (Haw. 1970)	§ 6.4
Jacksonville Port Auth. v. Parkhill-Goodloe Co., 362 So. 2d 1009 (Fla. 1978)	§ 4.9
Jahnke Serv., Inc. v. Department of Transp., 322 S.E.2d 505 (Ga. Ct. App. 1984)	§ 4.7
James Julian, Inc. v. President & Comm'r, 346 F.2d 205 (4th Cir. 1965)	§ 5.4
Jasper Constr. Co. v. Foothill Junior College Dist., 153 Cal. Rptr. 767 (Ct. App. 1979)	§ 6.4
J.E. Brenneman Co. v. Commonwealth Dep't of Transp., 424 A.2d 592 (Pa. Commw. Ct. 1981)	§§ 1.4, 4.7
J.E. Robertson Co. v. United States, 437 F.2d 1360 (Ct. Cl. 1971)	§ 8.3
Jefferson Constr. Co. v. United States, 364 F.2d 420 (Ct. Cl. 1966), *cert. denied,* 386 U.S. 914 (1967)	§ 2.4
Jen-Beck Assocs., VABCA No. 2121, 86-3 B.C.A. (CCH) ¶ 19,056 (1986)	§ 8.3
J.F. Shea Co., IBCA No. 1191-4-78, 82-1 B.C.A. (CCH) ¶ 15,705 (1982)	§§ 2.3, 2.4

Case	*Book §*
J.F. Shea Co. v. United States, 435 F.2d 873 (Ct. Cl. 1990)	§ 2.4
J.F. Shea Co. v. United States, 4 Cl. Ct. 46 (1983), *aff'd,* 754 F.2d 338 (Fed. Cir. 1985)	§§ 1.4, 2.1, 4.7, 4.8, 9.2, 9.10
J.L. Simmons Co. v. United States, 188 Ct. Cl. 684, 412 F.2d 1360 (1969)	§ 4.3
J.M.J. Invs., AGBCA Nos. 87-259-1, 87-341-1, 91-3 B.C.A. (CCH) ¶ 24,072 (1991)	§§ 8.2, 8.3
John A. Johnson Contracting Corp. v. United States, 132 F. Supp. 698 (Ct. Cl. 1955)	§ 9.7
John C. Grimberg Co., ASBCA Nos. 40670-72, 91-1 B.C.A. (CCH) ¶ 23,454 (1990)	§ 8.12
Joseph A. Cairone, Inc., ASBCA No. 20504, 81-2 B.C.A. (CCH) ¶ 15,220 (1981)	§§ 5.5, 8.8
Joseph F. Trionfo & Sons v. Board of Educ., 395 A.2d 1207 (Md. Ct. Spec. App. 1979)	§ 9.5
Joseph Morton Co., GSBCA No. 4611, 80-2 B.C.A. (CCH) ¶ 14,771 (1980)	§ 8.3
Joseph Penner, GSBCA No. 4647, 80-2 B.C.A. (CCH) ¶ 14,604 (1980)	§ 11.3
Josun, Inc., AGBCA Nos. 88-113-4, 82-249-4, 88-2 B.C.A. (CCH) ¶ 20,590 (1988)	§ 8.10
Kaiser Indus. Corp. v. United States, 340 F.2d 322 (Ct. Cl. 1965)	§§ 2.3, 2.4, 6.2, 8.3, 8.6
Kalisch-Jarcho v. City of N.Y., 448 N.E.2d 413 (N.Y. 1983)	§ 1.5
Kalisch-Jarcho, Inc. v. City of N.Y., 533 N.E.2d 258 (N.Y. 1988)	§ 13.6
Keang Nam Enters., Ltd., ASBCA No. 13747, 69-1 B.C.A. (CCH) ¶ 7705 (1969)	§ 8.3
Kenny Constr. Co. v. Metropolitan Sanitary Dist., 309 N.E.2d 221 (Ill. 1974)	§ 11.9
Kenny Constr. Co. v. Metropolitan Sanitary Dist., 288 N.E.2d 1 (Ill. 1971)	§ 3.3
Key Int'l Mfg. Inc. v. Morse/Diesel, Inc., 536 N.Y.S.2d 792 (App. Div. 1988)	§ 6.5
Kinetic Builders, Inc., ASBCA No. 32627, 88-2 B.C.A. (CCH) ¶ 20,657 (1988)	§§ 8.3, 8.10
Latham Co. v. United States, 20 Cl. Ct. 122 (1990)	§ 2.3
Lawrence-Lynch Corp. v. Department of Envt'l Management, 458 N.E.2d 332 (Mass. 1983)	§ 7.2
Layne v. GAF Corp., WL 155959 (Ohio Comm. Pl. 1989)	§ 13.5
Leal v. United States, 149 Ct. Cl. 451 (1960)	§ 8.8
Leiden Corp., ASBCA No. 26136, 83-2 B.C.A. (CCH) ¶ 16,612 (1983)	§§ 5.5, 8.9, 8.10
Leonard Blinderman Constr. Co., ASBCA No. 18946, 75-1 B.C.A. (CCH) ¶ 11,018 (1975)	§ 5.5
L.G. Everist, Inc., 231 Cl. Ct. 1013 (1982) *cert. denied,* 461 U.S. 957 (1983)	§ 8.6

Case	*Book §*
Moorhead Constr. Co. v. City of Grand Forks, 508 F.2d 1008 (8th Cir. 1975)	§ 3.3
Morrison-Knudsen Co. v. United States, 397 F.2d 826 (Ct. Cl. 1968)	§§ 1.4, 2.4, 9.2, 9.4
Morrison-Knudsen v. United States, 345 F.2d 535 (Ct. Cl. 1965)	§§ 5.4, 9.2, 9.4
Morrison-Knudsen Co. v. State, 519 P.2d 834 (Alaska 1974)	§ 6.2
Murphy v. City of Springfield, 738 S.W.2d 521 (Mo. Ct. App. 1987)	§ 4.9
Nelson Bros. Constr. Co. v. United States, 225 Ct. Cl. 703, 650 F.2d 290 (1980)	§ 8.3
Nelson Constr. Co. v. Port of Bremerton, 582 P.2d 511 (Wash. Ct. App.), *review denied,* 91 Wash. 2d 1002 (Wash. 1978)	§ 6.3
New Pueblo Constr. Co. v. State, 696 P.2d 185 (Ariz. 1985)	§ 8.3
Newberry Square Dev. Corp. v. Southern Landmark, Inc., 578 So. 2d 750 (Fla. Dist. Ct. App. 1991)	§ 1.5
Norair Eng'g Corp., GSBCA Nos. 2394, 2630, 72-1 B.C.A. (CCH) ¶ 9305 (1972)	§ 11.5
Norfolk Dredging Co. v. United States, 360 F.2d 619 (Ct. Cl.), *cert. denied,* 385 U.S. 919 (1966)	§ 9.7
North Harris County Junior College Dist. v. Fleetwood Constr. Co., 604 S.W.2d 247 (Tex. Civ. App. 1980) writ ref'd n.r.e.	§§ 3.1, 9.5
Northwest Painting Serv. Inc., ASBCA No. 27854, 84-2 B.C.A. (CCH) ¶ 17,474 (1984)	§ 8.9
Ogden Allied Servs. Corp. v. General Servs. Admin., GSBCA Nos. 9916, 10572, 1992 GSBCA LEXIS 200 (Jun. 4, 1992)	§ 11.3
Olshan Demolishing Co. v. Angleton Indep. Sch. Dist., 684 S.W.2d 179 (Tex. App. 1984) writ ref'd n.r.e.	§§ 3.1, 3.3
Oregon State Highway Comm'n v. DeLong Corp., 495 P.2d 1215 (Or. Ct. App. 1972), *cert. denied,* 411 U.S. 965 (1973)	§ 6.2
Overhead Elec. Co., ASBCA No. 22210, 78-2 B.C.A. (CCH) ¶ 13,440 (1978)	§ 8.9
Overland Elec. Co., ASBCA No. 454, 64-1 B.C.A. (CCH) ¶ 4359 (1964)	§ 9.7
Owners Realty Management Constr. Corp. v. Board of Educ., 554 N.Y.S.2d 648 (App. Div. 1990)	§ 13.6
Pacific Alaska Contractors, Inc. v. United States, 434 F.2d 461 (Ct. Cl. 1971)	§§ 2.3, 9.4
Passaic Valley Sewerage Comm'rs v. Holbrook Cabot & Rollins Corp., 6 F.2d 721 (3d Cir.), *cert. denied,* 269 U.S. 582, 46 S. Ct. 107 (1925)	§ 4.8

Case	*Book §*
Paul N. Howard Co. v. Puerto Rico Aqueduct Sewer Auth., 744 F.2d 880 (1st Cir. 1984)	§ 8.3
Pavement Specialists, Inc., ASBCA No. 17410, 73-2 B.C.A. (CCH) ¶ 10,082 (1973)	§ 5.5
P-B Eng'g Co., ASBCA Nos. 39041, 39042, 39043, 90-2 B.C.A. (CCH) ¶ 22,686 (1990)	§ 2.3
Pennsylvania Turnpike Comm'n v. Smith, 39 A.2d 139 (Pa. 1944), *overruled by,* 158 A.2d 461 (Pa. 1960)	§ 2.4
Perini Corp., ENGBCA No. 3745, 78-1 B.C.A. (CCH) ¶ 13,191 (1978)	§ 11.3
Perini Corp. v. United States, 381 F.2d 403 (Ct. Cl. 1967)	§§ 2.4, 11.9
Peter Kiewit Sons' Co. v. United States, 74 F. Supp. 165 (Ct. Cl. 1947)	§ 2.4
Petrochem Servs., Inc. v. United States, 837 F.2d 1076 (Fed. Cir. 1988)	§§ 11.3, 11.5
Pinner Constr. Co., PSBCA Nos. 2425, 2470, 2543, 91-3 B.C.A. (CCH) ¶ 24,143 (1991)	§ 8.3
P.J. Maffei Bldg. Wrecking Corp. v. United States, GSBCA No. 5031, 80-2 B.C.A. (CCH) ¶ 17,764 (1980), *aff'd,* 732 F.2d 913 (Fed. Cir. 1984)	§§ 1.4, 2.3, 6.3, 8.2, 8.3, 9.4, 9.5
Portable Rock Prod. Co. v. United States, 4 Cl. Ct. 495 (1984)	§ 6.3
Potomac Co., ASBCA No. 25371, 81-1 B.C.A. (CCH) ¶ 14,950 (1981)	§ 8.9
Praxis-Assurance Venture, ASBCA No. 24748, 81-1 B.C.A. (CCH) ¶ 15,028 (1981)	§§ 8.3, 8.9, 9.7
P.T.&L. Constr. Co. v. State, 531 A.2d 1330 (N.J. 1987)	§§ 1.5, 4.9, 6.1, 6.2, 6.3, 11.3
Public Constructors, Inc. v. State, 390 N.Y.S.2d 481 (App. Div. 1977)	§§ 6.3, 6.5
Puget Sound Bridge & Dredging Co. v. United States, 131 Ct. Cl. 490, 130 F. Supp. 368 (1955)	§ 9.2
Randall H. Sharpe, ASBCA No. 22800, 79-1 B.C.A. (CCH) ¶ 13,869 (1979)	§ 8.9
Ray D. Lowder, Inc. v. North Carolina State Highway Comm'n, 217 S.E.2d 682 (N.C. Ct. App.), *cert. denied,* 218 S.E.2d 467 (N.C. 1975)	§ 6.4
Ray W. Lynch, IBCA No. 764269, 69-2 B.C.A. (CCH) ¶ 8026 (1969)	§ 9.7
Raymond Int'l, Inc., ASBCA No. 13121, 70-1 B.C.A. (CCH) ¶ 8341 (1970)	§ 5.5
Richard Goettle, Inc. v. Tennessee Valley Auth., 600 F. Supp. 7 (N.D. Miss. 1984)	§ 8.3
R.L. Spencer Constr. Co., ASBCA No. 18450, 75-2 B.C.A. (CCH) ¶ 11,604 (1975)	§ 8.3
Roanoke Hosp. Ass'n v. Doyle & Russell, Inc., 214 S.E.2d 155 (Va. 1975)	§ 12.4

Case	*Book §*
Robert D. Carpenter, Inc., ASBCA No. 22297, 79-1 B.C.A. (CCH) ¶ 13,675 (1979)	§ 8.9
Robert E. McKee Gen. Constr., Inc. v. United States, ASBCA No. 5521, 60-1 B.C.A. (CCH) ¶ 2526 (1960)	§ 8.3
Robert E. McKee, Inc. v. City of Atlanta, 414 F. Supp. 957 (N.D. Ga. 1976)	§§ 4.9, 6.4
Roberts Constr. Co., ASBCA No. 33949, 87-2 B.C.A. (CCH) ¶ 19,897 (1987)	§ 2.3
Rocell Constr. Co. v. State, 141 N.Y.S.2d 463 (Ct. Cl. 1955)	§ 6.3
Roen Salvage Co., Appeal of, ENGBCA No. 36701, 79-2 B.C.A. (CCH) ¶ 13,882 (1979)	§ 9.7
Roger J. Au & Son, Inc., IBCA No. 1303979, 84-1 B.C.A. (CCH) ¶ 17,094 (1984)	§ 9.6
R.T. Crowley, Inc., GSBCA No. 11080 (9521)-REIN, 92-1 B.C.A. (CCH) ¶ 24,499 (1991)	§ 11.3
S. Kane & Sons, Inc., VACAB No. 1254, 78-1 B.C.A. (CCH) ¶ 13,100 (1978)	§ 11.9
Sabella v. Wisler, 377 P.2d 889 (Cal. 1963)	§ 6.5
Salinas, City of v. Souza & McCue Constr. Co., 424 P.2d 921 (Cal. 1967)	§ 6.2
Saltany v. Reagan, 886 F.2d 438 (D.C. Cir. 1989), *cert. denied,* 110 S. Ct. 2172 (1990)	§ 6.2
Sanders Constr. Co., 220 Ct. Cl. 639 (1979), *rev'g* ENGBCA No. 3482, 75-2 B.C.A. (CCH) ¶ 11,481 (1975)	§ 11.5
Sandwich Islands Constr. Inc., ASBCA No. 35244, 88-3 B.C.A. (CCH) ¶ 21,143 (1988)	§§ 5.3, 8.8
Sasso Contracting Co. v. State, 414 A.2d 603 (N.J. Super. Ct. App. Div.), *cert. denied,* 425 A.2d 265 (N.J. 1980)	§§ 6.1, 9.5
Saturn Constr. Co., ASBCA No. 22653, 82-1 B.C.A. (CCH) ¶ 15,704 (1982)	§ 11.3
Schnip Bldg. Co. v. United States, 645 F.2d 950 (Ct. Cl. 1981)	§§ 9.6, 11.3
Schouten Constr. Co., DOT CAB No. 77-4, 79-1 B.C.A. (CCH) ¶ 13,630 (1978)	§ 11.3
Schutt Constr. Co. v. United States, 353 F.2d 1018 (Ct. Cl. 1965)	§ 2.4
Seiler v. Ostarly, 525 So. 2d 1207 (La. Ct. App. 1988)	§ 6.5
Servidone Constr. Corp. v. United States, 19 Cl. Ct. 346 (1990), *aff'd,* 931 F.2d 860 (Fed. Cir. 1991)	§§ 2.1, 2.3, 2.4
Shank-Artukovich v. United States, 13 Cl. Ct. 346 (1987), *aff'd,* 848 F.2d 1245 (Fed. Cir. 1988)	§§ 2.4, 11.1
Sharjah Inv. Co. (UK) Ltd. v. P.C. Telemart, Inc., 108 F.R.D. 90 (S.D.N.Y. 1985)	§ 6.2
Shepherd v. United States, 113 F. Supp. 648 (Ct. Cl. 1953)	§ 3.3
Shumate Constructors, Inc., VABCA No. 2772, 90-3 B.C.A. (CCH) ¶ 22,946 (1990)	§§ 8.2, 8.3, 8.4, 8.8, 8.9

Case	*Book §*
Sierra Blanca, Inc., ASBCA Nos. 30943, 32286, 34838, 91-2 B.C.A. (CCH) ¶ 23,990 (1991)	§ 8.3
Simpson Constr. Co., VABCA No. 3176, 91-1 B.C.A. (CCH) ¶ 23,630 (1990)	§§ 5.2, 5.4, 8.2, 8.3, 8.6
Simpson Timber Co. v. Palmberg Constr. Co., 377 F.2d 380 (9th Cir. 1967)	§ 6.2
Singleton Contracting Corp., IBCA No. 14131280, 81-2 B.C.A. (CCH) ¶ 15,269 (1981)	§ 9.6
S.J. Groves & Sons Co. v. State, 273 S.E.2d 465 (N.C. Ct. App. 1980), *reh'g denied,* 279 S.E.2d 353 (N.C. 1981)	§§ 3.3, 9.10
Skip Kirchdorfer, Inc., ASBCA No. 23443, 79-2 B.C.A. (CCH) ¶ 14,092 (1979)	§ 8.2
Skopek Bros., Inc. v. Webster Hous. Auth., 416 N.E.2d 1006 (Mass. App.Ct. 1981)	§§ 1.6, 3.4
S&M Constructors v. City of Columbus, 434 N.E.2d 1349 (Ohio 1982)	§§ 4.7, 6.6
S.O.G. of Ark. v. United States, 546 F.2d 367 (Ct. Cl. 1976)	§ 8.3
Southern Paving Corp., AGBCA No. 741031, 77-2 B.C.A. (CCH) ¶ 12,813 (1977)	§ 9.7
Southwestern Eng'g v. Cajun Elec. Power, 915 F.2d 972 (5th Cir. 1990)	§ 1.5
Souza & McCue Constr. Co. v. Superior Court, 370 P.2d 338 (Cal. 1962)	§ 6.2
Spirit Leveling Contractors v. United States, 19 Cl. Ct. 84 (1989)	§ 6.3
S.S. Mullen, Inc., ASBCA No. 8583, 65-1 B.C.A. (CCH) ¶ 4644 (1965)	§ 9.7
State Mechanical Corp., VABCA No. 2797, 91-2 B.C.A. (CCH) ¶ 23,830 (1991)	§ 11.9
Stauffer Constr. Co. v. Board of Educ., 460 A.2d 609 (Md. Ct. Spec. App. 1983)	§ 9.6
Steinberg v. Fleisher, 706 S.W.2d 901 (Mo. Ct. App. 1986)	§§ 6.2, 6.6
Stewart v. Cox, 362 P.2d 345 (Cal. 1961)	§ 6.5
Stock & Grove, Inc. v. United States, 493 F.2d 629 (Ct. Cl. 1974)	§§ 2.3, 8.4, 11.3
Structural Preservation Sys., Inc., 3 MSBCA ¶ 234 (1989)	§ 9.2
Stuyvesant Dredging Co. v. United States, 834 F.2d 1576 (Fed. Cir. 1987)	§§ 4.3, 6.4
Tanglewood E. Homeowners v. Charles-Thomas, Inc., 849 F.2d 1568 (5th Cir. 1988)	§ 2.4
Thorn Constr. v. Utah Dep't of Transp., 598 P.2d 365 (Utah 1979)	§ 4.5
Thrush v. Morrison, 665 F. Supp. 372 (M.D. Pa. 1987), *aff'd,* 845 F.2d 1016 (3d Cir.), *cert. denied,* 109 S. Ct. 74 (1988)	§ 6.2
Titan Midwest Constr. Corp., ASBCA No. 23594, 81-1 B.C.A. (CCH) ¶ 15,067 (1981)	§ 8.9
Tobin Quarries, Inc. v. United States, 114 Ct. Cl. 286, 84 F. Supp. 1021 (1949)	§ 8.6

Case	*Book §*
Tombigbee Constructors v. United States, 420 F.2d 1037 (Ct. Cl. 1970)	§ 2.3
Tri-Messine Constr. Co., PSBCA No. 312, 78-1 B.C.A. (CCH) ¶ 13,023 (1978)	§ 11.3
Turnkey Enters., Inc. v. United States, 597 F.2d 750 (Ct. Cl. 1979)	§§ 1.4, 6.4, 9.7
Tuscaloosa City Bd. of Educ. v. American/Owens, Inc., 486 So. 2d 405 (Ala. 1986)	§ 13.6
Tutor-Saliba, ASBCA No. 23776, 79-2 B.C.A. (CCH) ¶ 14,137 (1979)	§§ 5.5, 8.10
Umpqua River Navigation Co. v. Crescent City Harbor Dist., 618 F.2d 588 (9th Cir. 1980)	§§ 4.7, 4.9, 9.4
Unitec, Inc., DOT BCA No. 7729, 78-1 B.C.A. (CCH) ¶ 12,958, *aff'd,* 78-2 B.C.A. (CCH) ¶ 13,222 (1978)	§ 9.7
United Contractors v. United States, 368 F.2d 585 (Ct. Cl. 1966)	§§ 2.4, 8.3, 11.3
United States v. Atlantic Dredging Co., 253 U.S. 1, 40 S. Ct. 423 (1920)	§ 6.4
United States v. Joseph A. Holpuch Co., 328 U.S. 234 (1946)	§ 8.1
United States v. Spearin, 248 U.S. 132, 39 S. Ct. 59 (1918)	§§ 4.2, 4.3, 5.1, 6.4
Universal Contracting & Brick Pointing Co. v. United States, 19 Cl. Ct. 785 (1990)	§§ 4.3, 13.6
Upjohn Co. v. New Hampshire Ins. Co., Nos. 86906–86908 (Mich. 1991)	§ 12.12
Utility Contractors, Inc. v. United States, 8 Cl. Ct. 42 (1985), *aff'd,* 709 F.2d 90 (Fed Cir.), *cert. denied,* 479 U.S. 827, 107 S. Ct. 104 (1986)	§§ 4.3, 9.7
Utley-James, Inc. v. United States, 14 Cl. Ct. 804 (1988), *aff'g* GSBCA No. 5370, 85-1 B.C.A. (CCH) ¶ 17,816 (1984)	§ 11.9
Valentini v. City of Adrian, 79 N.W.2d 885 (Mich. 1956)	§ 6.3
Vann v. United States, 420 F.2d 968 (Ct. Cl. 1970)	§ 2.3
Village Dev. Co. v. Filice, 526 P.2d 83 (Nev. 1974)	§§ 6.1, 6.5
Volt Info. Sciences v. Board of Trustees, 489 U.S. 468, 109 S. Ct. 1248 (1989)	§ 1.6
Waggoner v. Midwestern Dev. Inc., 154 N.W.2d 803 (S.D. 1967)	§§ 6.2, 6.3, 6.5
Warner Constr. Corp. v. City of L.A., 466 P.2d 996 (Cal. 1970)	§§ 1.5, 6.2, 6.4
Warner Elec., Inc., VABCA No. 2106, 85-2 B.C.A. (CCH) ¶ 18,131 (1985)	§ 2.3
Warren Bros. Rds. Co. v. United States, 123 Ct. Cl. 48, 105 F. Supp. 826 (1952)	§ 9.7
Warren Painting Co., ASBCA No. 18456, 74-2 B.C.A. (CCH) ¶ 10,834 (1974)	§§ 8.3, 9.7, 11.3

INDEX